Michael Giebels

Arctic Permafrost and Greenhouse Gases

Michael Giebels

Arctic Permafrost and Greenhouse Gases

Methane Fluxes in Wet Polygonal Tundra on
Samoylov Island, Lena River Delta, Siberia

VDM Verlag Dr. Müller

Imprint

Bibliographic information by the German National Library: The German National Library lists this publication at the German National Bibliography; detailed bibliographic information is available on the Internet at http://dnb.d-nb.de.

Cover image: www.purestockx.com

Publisher:
VDM Verlag Dr. Müller Aktiengesellschaft & Co. KG, Dudweiler Landstr. 125 a, 66123 Saarbrücken, Germany,
Phone +49 681 9100-698, Fax +49 681 9100-988,
Email: info@vdm-verlag.de

Produced in USA and UK by:
Lightning Source Inc., La Vergne, Tennessee, USA
Lightning Source UK Ltd., Milton Keynes, UK
BookSurge LLC, 5341 Dorchester Road, Suite 16, North Charleston, SC 29418, USA

ISBN: 978-3-8364-8831-0

Contents

I   Abstract

Methane (CH_4) is an important radiatively active trace gas. As its atmospheric concentration has more than doubled since pre-industrial times, a need to understand the processes involved driving CH_4 exchange between lithosphere and atmosphere exists, especially in the context of climate change, where the Arctic is of major interest. This study aims at providing spatially highly resolved small-scale methane flux measurements of the Arctic wet polygonal tundra on Samoylov Island in the Lena River Delta region of North Siberia. Small-scale process understanding is needed to fully understand and accurately interpret ecosystem-scale measurements provided by the eddy covariance method, where the focus is primarily put on high temporal resolution. For this purpose, an extensive array of micro-sites for plot-based CH_4 flux measurements sampled on an almost daily basis from July through September 2006. CH_4 flux measurements were conducted using the closed chamber measurement method. The data set presented in this study is unique with regard to its extend, the investigation area, and it's further potential for integrating flux measurements on different spatial scales. The investigation area in the centre of Samoylov Island was a source of methane in the observed summer season, but not as strong a source as might be expected from a "wet" tundra. This, along with the conclusion, that methane fluxes differed by an order of magnitude between polygons with low-centre character and polygons with high-centre character or polygon rims, respectively, is seen as an indicator for the high spatial variability of the methane emissions. Using simple multiple linear regression model approaches, which successfully described measured methane fluxes from wet polygon centres, soil surface temperature T_{Soil-1} was identified as a main environmental control. Other identified parameter, which varied between sites, raise questions regarding the reliability of closed chamber methods (e.g. where fluxes were dependent on temperature inside the chamber), or suggested an additional focus for future investigation (e.g. into the importance of the water regime and hydrology).

II   Zusammenfassung

Methan (CH$_4$) ist ein wichtiges Treibhausgas. Seine atmosphärische Konzentration hat sich seit dem vorindustriellen Zeitalter mehr als verdoppelt. Somit besteht ein erhöhter Bedarf diejenigen Prozesse zu verstehen, die am Austausch von Methan besonders im Hinblick auf die globale Erwärmung, mit der Lithosphäre und der Atmosphäre beteiligt sind. Hierbei ist die arktische Region von besonderem Interesse. Diese Arbeit unternimmt den Versuch, räumlich hoch aufgelöste kleinskalige Methanflussmessungen arktischer feuchtpolygonaler Tundraböden auf der Insel Samoylov im nordsibirischen Lenadelta zu bestimmen. Prozessverständnis auf dieser Ebene ist notwendig, um Messungen auf der Ökosystem-Ebene der Eddy-Kovarianz-Methode vollständig zu verstehen und interpretieren zu können, deren Fokus hauptsächlich auf der hohen zeitlichen Auflösung liegt. Zu diesem Zweck wurde ein umfangreicher Versuch mit verschiedenen Kleinräumen zur Methanflussmessungen im Zeitraum von Juli bis September 2006 durchgeführt. Die Methanflussmessungen wurden mit geschlossenen Hauben durchgeführt. Der Datensatz, der dieser Arbeit zu Grunde liegt, ist in seiner Art einzigartig in Umfang, Größe des Untersuchungsgebietes und seinem Potential für das Verständnis der Flussmessungen auf unterschiedlichen räumlichen Skalen. Das Untersuchungsgebiet im Zentrum der Insel Samoylov stellte sich für den Zeitraum der untersuchten Sommersaison als eine Quelle atmosphärischen Methans dar, jedoch nicht in dem für eine „feuchte" Tundra zu erwarten gewesenen Maße. Dies und die Erkenntnis, dass sich die Methanflüsse von Polygonen unterschiedlicher Genese teilweise erheblich unterschieden, ist ein Zeiger für die hohe räumliche Variabilität der Methanemissionen. Mit Hilfe einfacher multipler linearer Regressionsmodelle, die gemessene Methanflüsse von feuchten Polygonzentren erfolgreich wiedergeben konnten, konnte die Temperatur unmittelbar unter der Bodenoberfläche als ein bedeutender steuernder Umweltparameter identifiziert werden. Weitere identifizierte Parameter, die sich je nach Untersuchungsraum unterschieden, werfen Fragen auf, ob die Methodik der geschlossenen Hauben für die Methanflussmessung geeignet erscheint (Flüsse hingen von der Temperatur innerhalb der geschlossenen Hauben ab), beziehungsweise inwieweit bisher nicht beachtet Umweltparameter in zukünftigen Methanflussbetrachtungen in erhöhte Beachtung verdienen (beispielsweise die Bedeutung der individuellen Wasserregime und ihre Hydrologie).

III Danksagungen

Diese Arbeit entstand in Kooperation des Instituts für Geoökologie der Technischen Universität Braunschweig mit dem Alfred-Wegener-Institut (AWI) für Polar- und Meeresforschung in Potsdam.

In diesem Zusammenhang möchte ich mich bei Herrn Prof. Dr. Wolfgang Durner für die Möglichkeit bedanken, eine externe Diplomarbeit schreiben zu dürfen. Des Weiteren muss ich mich für die mir dafür entgegenbrachte Geduld und Rücksichtnahme bedanken. Ich danke Herrn Prof. Dr. Hans-Wolfgang Hubberten als Leiter des AWI Potsdam stellvertretend für alle Beteiligten, die meine Teilnahme an der AWI-Expedition ins Lenadelta im Sommer letzten Jahres ermöglicht haben. Hier ist noch Herr Dr. Dirk Wagner zu erwähnen, der im Vorfeld der Expedition die Koordination und Korrespondenz führte. Nicht zuletzt muss ich mich bei ihm und Herrn Sascha Iden vom Institut für Geoökologie für die geleistete Betreuungsarbeit bedanken.

Recht herzlich möchte ich mich bei allen Teilnehmern der „System Laptev Sea Lena 2006" Expedition bedanken: bei Günter „Molo" Stoof für zahllose Eindrücke, Tipps zum Überleben auf der Trauminsel und den Kakao auf dem Berliner Alexanderplatz um 4 Uhr morgens. Bei Waldemar Schneider für die logistische Organisation der Expedition. Bei Merten Minke für sein umfassendes Wissen Moose, Flechten und Gräser betreffend und seinen unbezähmbaren Forscherdrang. Bei Stefanie Kirschke für das selbstlose Überlassen eines köstlichen Nudossi-Glases. Bei Sergei Volkov, dem alten Taxifahrer, und seiner Frau Olga Volkova als den eigentlichen Bewohnern der Trauminsel. Bei allen anderen Mitbewohnern und -forschern für die gesammelten Erfahrungen im kollegialen wie im zwischenmenschlichen Sinne: Jürgen „Rebyonok" Joseph, Dr. Julia Boike, Konstanze „009" Piel, Simone „Owl-Hunter" Bircher, und nicht zu vergessen die russische Forscherseite: Irina „Ira" Wischnjakova, Anastasia „Nastia" Germogenova, Dr. Katya Abramova mit Anya, Dr. Dmitry Bolschiyanov und Aleksander Makarov vom AARI St. Petersburg sowie den russischen Partnern des AWI Potsdam Alexander Gukov vom Lena Delta Reservat in Tiksi, Mikhail Grigoriev vom Permafrost Institute in Yakutsk, Alexander Derevyagin von der Moscow State University, und Dmitri Melnitschenko von der Hydro Base in Tiksi.

Ein herzliches Dankeschön geht auch an Herrn Dr. Lars Kutzbach von der Universität Greifswald für die freundliche Unterstützung bei der Flussberechnung sowie die nette familiäre Unterbringung beim Besuch in Greifswald und nicht zuletzt auch für seine Last-Minute-Revision über ein ländliches Telefonkabel. Ich danke Frau Dr. Dagmar Söndgerath vom Institut für Geoökologie für ihre spontane Statistik-Nachhilfe und Christian Wille von der Universität Greifswald für seine Revisionstätigkeit in letzter Minute.

Mein allergrößter Dank gilt Torsten Sachs vom AWI Potsdam. Über ihn entstand der Kontakt zu dieser Arbeit. Ich bedanke mich an dieser Stelle für die reichhaltige und kaum wieder gutzumachende Unterstützung: Für die mehrfache Unterbringung in Potsdam, Ratschläge, Tipps, Anmerkungen, Revisionsbeiträge zu jeder Tages- und Nachtzeit, aber insbesondere für sein unerschöpfliches Maß an Geduld, Verständnis und Rücksichtnahme. Im vergangenen Jahr ist dabei eine neue Freundschaft entstanden.

Für Geduld, Verständnis und Rücksichtnahme in den letzten Monaten ist noch vielen weiteren zu danken. Allen voran meinem Vater, meiner Mutter und meiner Schwester für die dennoch nicht ausgebliebene Unterstützung, Fürsorge und Anteilnahme. Ich bedanke mich auch bei allen geduldigen und verständnisvollen Freunden, sowie speziell bei Anna Behrendt, Martin Strelow, Sabrina Gärtner und Tanja Müller für ihre Revisionsbeiträge, nützlichen Formatierungstipps und Literaturakquise.

Ich danke Nicole für ihre wirklich grenzenlose Geduld, Toleranz und Liebe.

Michael Giebels, 04. September 2007

I V   List of figures

V   List of tables

VI   List of mathematical formula symbols

Symbol	Description
$cumF$ (g m^{-2} d^{-1})	cumulative flux
D	diffusion coefficient
F (g m^{-2} d^{-1})	vertical flux
k_i (m/s)	transfer velocity
L (cm)	depth of active layer
p (kPa)	barometric pressure
p (Pa)	partial pressure
PR (mm)	precipitation
R^2	adjusted coefficient of determination
$\triangle G°$ (kJ)	free standard enthalpy
T (°C)	temperature
T (K)	temperature
t (min)	time
T_{air} (°C)	air temperature
T_{ch} (°C)	chamber temperature
T_{soil} (°C)	soil temperature
u_* (m/s)	friction velocity
W (cm)	Water level
W_{col} (cm)	Water column
θ (vol %)	soil moisture

VII List of abbreviations, units and emp. Formulas

a.s.l.	above sea level
AG	Aktiengesellschaft, public limited company
approx.	approximately
AWI	Alfred-Wegener-Institue for Polar and Marine Research
c	concentration
CFCs	halocarbon gases
ch	chamber
CH_3COOH	acetic acid
CH_3OH	methanol
CH_4	methane
cl	clear
cm	centimetre
cmp.	compare
CO_2	carbon dioxide
col	column
d	day
E	East
e.g.	exempli gratia, for example
emp.	empirical
eq.	equation
eqs.	equations
et al.	et alii, and others
g	gram
GmbH	Gesellschaft mit beschränkter Haftung, private limited company
H	hydrogen
H_2	hydrogen gas
H_2O	water
HCHO	formaldehyde
HCOOH	formic acid, formate
i.e.	id est, that is
IR	infrared
J	Joule
K	Kelvin
kJ	kiloJoule
km	kilometre
kPa	kilopascal
Ltd.	Limited
m	metre
mg	milligram
min	minute
mm	millimetre

mol	mol
mV	millivolt
N	North
N_2O	nitrous oxide
NAD^+	nicotinamide adenine dinucleotide, cell coenzyme
$NADH_2$	nicotinamide, niacin, vitamin B_3
NEP	net ecosystem production
nm	nanometre
no.	number
O_2	oxygen
O_3	ozone
op	opaque
p	barometric pressure
p	partial pressure
Pa	Pascal
PAR	Photosynthetically Active Radiation
ppm	parts per million
ppmv	volumetric concentration ppm
PR	precipitation
R	ideal gas constant
R^2	adjusted coefficient of determination
s	second
T	temperature
t	time
V	volume
%	percent
&	and
°C	degree celsius
…	range of a period or quantity

1   Introduction and objectives

Methane (CH_4) along with carbon dioxide (CO_2), water vapour (H_2O), nitrous oxide (N_2O), halocarbon gases (CFCs), and ozone (O_3) is an important radiatively active trace gas. These atmospheric gases let shortwave solar radiation pass through the atmosphere. The earth surface then absorbs shortwave radiation and transforms it into longwave thermal radiation, which is emitted back into the atmosphere but partially prevented from passing back into space by the trace gases mentioned above. This effect known as natural greenhouse effect causes a global average air temperature at ground level of +15°C instead of -18°C without it [*IPCC, 1990*], as a consequence from Kirchhoff's and Plank's radiation laws. However, the atmospheric composition of greenhouse gases has changed substantially. This leads to a change in global average air temperature [*IPCC, 2001*]. Despite the fact that water vapour is the most common greenhouse gas with a big reservoir represented in the world's oceans, it does not get much attention concerning its global warming potential. The main focus, especially in politics, currently rests on carbon gases, the concentration of which has changed significantly since the start of the industrial revolution. For example, the tropospheric concentration of CO_2 rose from 280ppm to 367ppm in the past two centuries while the CH_4 concentration doubled in the same time [*Etheridge et al., 1998*]. Another reason for the focus on carbon atmospheric gases is their effectiveness regarding their warming potential due to their physical characteristics.

With regard to the discussion whether current global warming is caused by anthropogenic activity [*Hasselmann et al., 2003; Rahmstorf et al., 2004*] or natural climate fluctuation [*von Storch et al., 2004; Moberg et al., 2005*], it is important to extend the knowledge about the interactions between terrestrial or marine activities and the atmosphere. Within this context the Arctic is of major interest for several reasons: It is observed to warm more rapidly and to a greater extent than the rest of the earth surface [*Serreze et al., 2000; Polyakov et al., 2003*], plus, it is an important area controlling many global processes, such as the atmospheric and oceanic circulations [*Stockner and Schmittner, 1997; Wood et al., 1999; Eugster et al., 2000*] and the regulation of the global greenhouse gas budgets [*Gorham, 1991; Roulet et al., 1992; Tenhunen, 1996*]. Mainly, arctic land surfaces are covered by tundra. Together with alpine tundra, arctic tundra covers 7.4% of the northern hemisphere's land area [*Matthews, 1983; Loveland et al.; 2002*].

The term tundra has its origin in the Sami language Kildin. It is defined as almost treeless area or as area where tree growth is hindered by very low temperatures and thus short growing seasons. Typical tundra vegetation are crouching shrubs, mosses, lichens and herbaceous plants such as sedges and grasses.

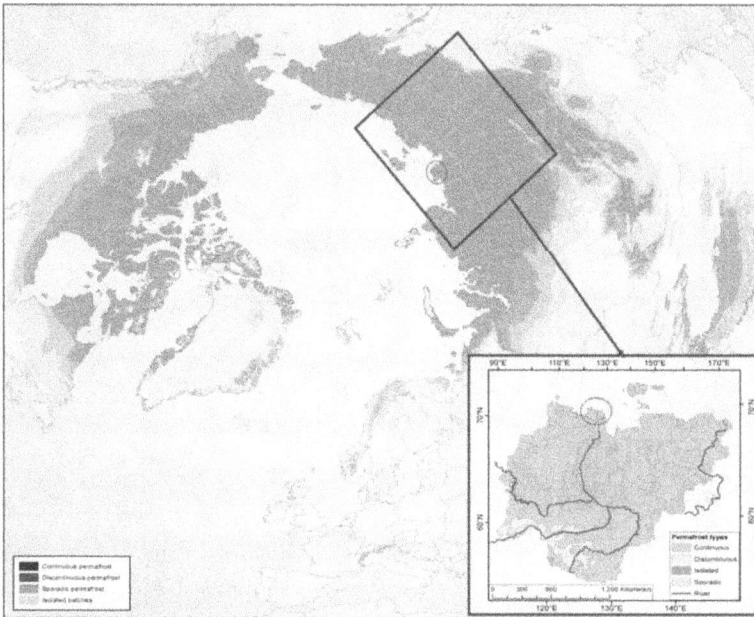

Figure 1-1:   Circumpolar distribution of permafrost [from *Parsons and Zhang, 2003*, modified by *Schnelle, 2007*] – the detail shows the permafrost distribution of Yakutia, encircled the Lena River Delta region.

Tundra soil is underlain by permafrost, which covers about 24% of the northern hemisphere's exposed land surface [*Zhang et al., 1999*]. *Van Everdingen* [*1998*] defines permafrost as soil or rock that remains below 0 °C for at least two consecutive years. Figure 1-1 shows the distribution of permafrost in the northern hemisphere according to *Parsons and Zhang* [*2003*]. In general, the thickness of permafrost increases polewards from some meters at the southern permafrost boundary to more than 1000m. The maximum depth of permafrost is reached in northwest Yakutia, central Siberia, with a depth of about 1500m [*Embelton & King, 1975*]. Due to seasonal surface thawing permafrost is covered by an unfrozen zone during summer months, called active layer. Depending on several parameters, e.g. the geographic latitude, exposition and substrate, active layers can reach a depth of 1m [*Schultz, 1995*]. Usually active layers in northern Siberia are not deeper than 0.5m [*Wein, 1999*].

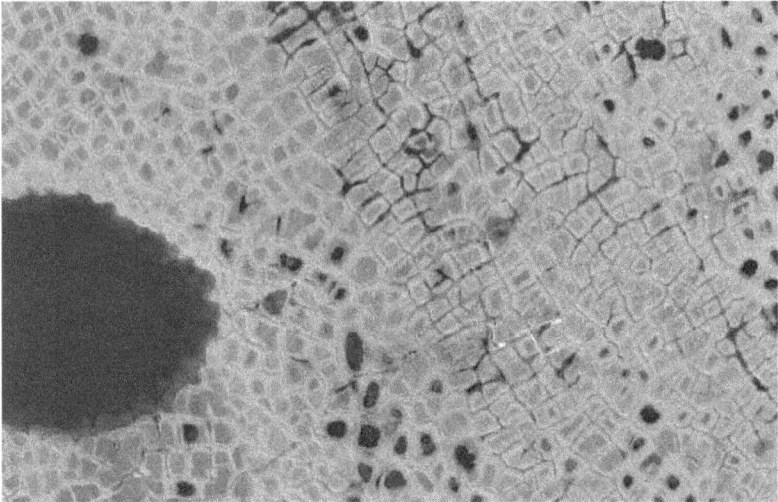

Figure 1-2: **Wet polygonal tundra on Samoylov Island** – dark blue coloured areas indicate open water in water-filled polygons, thermokarst cracks, and thermokarst lakes, respectively (Photograph by courtesy of Julia Boike, AWI Potsdam).

Typical forms of continuous permafrost soils are ice wedge polygons [*Schultz, 1995*]. Their genesis is hard to comprehend. Interrelations and interactions between relief difference, water saturation, substrate and temperature can result in completely different polygon development at the same site. Generally, ice wedge polygons are caused by the contraction of soil, forming cracks, which are filled with melt water during spring. The water refreezes immediately, forming small venes of ice in the soil. The continuous repetition of this seasonal cycle of contraction, thawing and freezing over long time periods leads to the development of ice wedge polygons. An area formed by ice wedges is characterized by a polygonal pattern as can be seen in Figure 1-2. Due to thermal expansion during summer, soil layers get elevated above ice wedges. Polygons with elevated rims and depressed centres are called "low-centre" polygons (Figure 1-3). The depressed centres are usually poorly drained and filled with water, which may result in initial thermokarst. If ice wedges thaw, formerly separated polygon centres can form larger polygon lakes (Figure 1-2). Another possible development is the so-called "high-centre" polygon (Figure 1-3): Thermokarst increases the thawing depth of the active layer and melting of the ice wedges at the surface. This may result in a better draining of the polygon centre. The rims flatten and the peaty centre grows up, similar to the genesis of a raised bog, until it is elevated above the surrounding area [*French, 1996* cited in *Schnelle, 2007*].

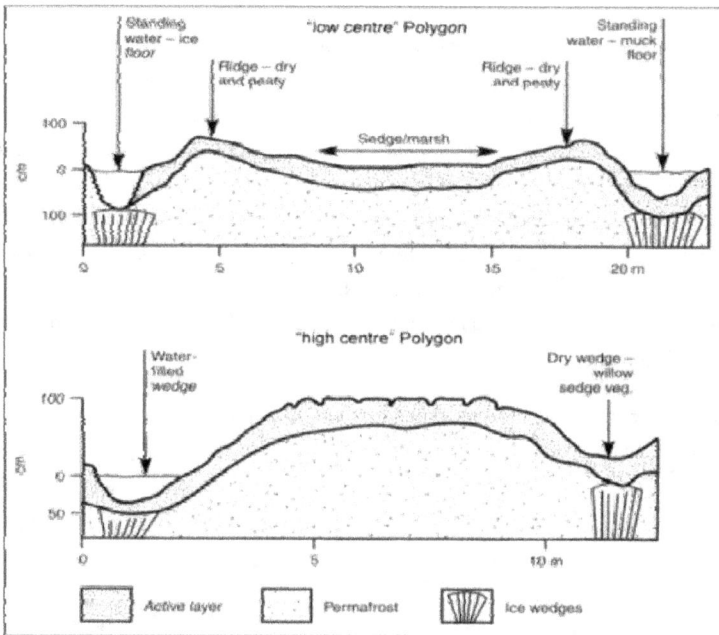

Figure 1-3: **Relief forms of low centre and high centre polygons** [*French, 1996*; modified by *Schnelle, 2007*].

Arctic tundra ecosystems have been major carbon sinks throughout historical and geological periods of time, because of their subzero characteristics. Decomposition of organic material is slowed down by the cold temperatures, the short growing season during Polar Day, overlain litter material, and the saturated soil water conditions due to the presence of permafrost [*Ping et al., 1998*]. However, in present time, more recent studies show that some arctic ecosystems change into carbon sources due to global warming. Because of the environmental conditions, arctic soils contain large amounts of organic carbon - at least 14% of the global soil organic carbon is stored in the tundra [*Post et al., 1982; Billings, 1987*]. A change of arctic tundra ecosystems from carbon sinks to carbon sources, due to the increased level of permafrost thawing caused by its susceptibility to global warming, can significantly increase the content of atmospheric carbon, either in the form of carbon dioxide (CO_2) or methane (CH_4) [*Oechel et al., 1993; Christensen, 1993; Zimov et al., 1997*]. For a better understanding of the dynamics of carbon exchange between tundra ecosystem and the atmosphere, detailed process-studies on multiple temporal and spatial scales are necessary.

For this purpose, in 2002 the Alfred Wegener Institute (AWI) in Potsdam established a seasonally operated eddy covariance system in the Lena River Delta in Northern Siberia, Russia. The system detects turbulent fluxes of momentum, sensible and latent heat, water vapour (H_2O), methane (CH_4), and carbon dioxide (CO_2) in the atmosphere near ground level with high temporal resolution. Additionally, various meteorological and soil climatic parameters are recorded. However, the eddy covariance system measures integrated net fluxes between soil/vegetation and the atmosphere (e.g. NEP = net ecosystem production) over a larger area of hectares to square kilometres, depending on measurement height and other conditions.

In the case of methane, the patterned ground of the polygonal tundra results in an extremely high small-scale spatial variability of emissions, which are up to an order of magnitude larger in polygon centres than on drier polygon rims. This spatial variability can not be resolved by the eddy covariance method but is crucial for a detailed process-understanding as well as in the context of up-scaling emissions from small, plot-based measurements to the landscape and regional scales.

This study aims at providing the spatially highly resolved small-scale methane flux measurements as well as the small-scale process understanding needed to fully understand and accurately interpret ecosystem-scale measurements provided by the eddy covariance method. For this purpose, an extensive array of micro-sites for plot-based flux measurements was installed during the AWI expedition Lena 2005 and sampled on an almost daily basis from July through September 2006.

2   Basics

In general, the exchange of matter or energy between atmosphere and lithosphere is referred to as flux. This study investigates the exchange of methane between different arctic tundra soil-vegetation units and the atmosphere. The sign convention in this study is such, that emission flux from lithosphere to atmosphere is noted as positive flux, while flux directed from the atmosphere into the lithosphere is noted as negative flux.

A basic method of measuring flux is the closed chamber method. The change of concentration of the investigated gas in a deployed chamber headspace over time is caused by the net exchange flux $F_{net}(t)$:

$$F_{net}(t) \; = \; \frac{dc}{dt}(t) \, \frac{p \, V}{R \, T \, A} \tag{1}$$

where c is the concentration, p is atmospheric pressure, V is the headspace volume, R is the ideal gas constant, T is temperature, and A is the chamber basal area.

Methane production in soil is mainly controlled by methanogenic microorganisms (*Archae*), which convert organic material into methane under anaerobic conditions (Table 2-1). Under anaerobic conditions biogenic methanogenesis by micro-organisms requires a redoxpotential of -150 to -300 mV, absence of nitrate and sulphate and presence of convertible organic material [*Heyer, 1990; Wang et al., 1993* cited in *Kutzbach, 2000*]. Such an environment is given in water saturated soils e.g. *Typic Historthels*. There, oxygen is respired by microbes and can be resupplied just in small amounts from atmosphere [*Scheffer and Schachtschabel, 1998*]. Even in partially drained and ventilated soils anaerobic microhabitats can exist, in which methane production is possible. Table 2-1 shows the most important reactions in the metabolisms of methanogens.

Table 2-1: Reactions in metabolisms of methanogenic microorganisms
[*Heyer, 1990*]

substrate input	metabolism reaction	energy balance
H_2/CO_2	$4\ H_2 + CO_2 \rightarrow CH_4 + 2\ H_2O$	$\Delta G^0 = -138.8$ kJ
formate	$4\ HCOOH \rightarrow CH_4 + 3\ CO_2 + 2\ H_2O$	$\Delta G^0 = -119.5$ kJ
acetate	$CH_3COOH \rightarrow CH_4 + CO_2$	$\Delta G^0 = -27.6$ kJ
methanol	$4\ CH_3OH \rightarrow 3\ CH_4 + CO_2 + 2\ H_2O$	$\Delta G^0 = -310.5$ kJ

On the other hand, methanotrophic microbes (*bacteria*) can degrade methane into carbon dioxide under aerobic conditions. Methane degradation is known as methane oxidation and takes place in a stepwise manner (Figure 2-1). Main oxidation activity takes place directly at the aerobic-anaerobic boundary, where both methane and oxygen are supplied [*Knoblauch, 1994; Whalen et al., 1996* cited in *Kutzbach, 2000*]. Up to 95% of the methane produced under anaerobic conditions can be degraded by methanotrophic bacteria on its way to the atmosphere [Holzapfel-Pschorn et al., 1986; King, 1990; Happell et al., 1993 cited in Kutzbach, 2000]. The amount of methane degradation is directly depending on its transport to the surface. The longer the time period methane is exposed to aerobic conditions, the higher is the likelihood of its degradation [Conrad, 1989 cited in Kutzbach, 2000].

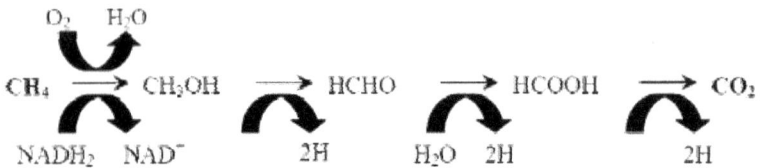

Figure 2-1: Methane oxidation by methanotrophic bacteria [Fritsche, 2002]

There are several ways of how soil-originated methane can reach the atmosphere (Figure 2-2). Convective gas transport in soil can be caused by pressure differences, which can appear due to changing water levels or strong winds. Usually, this transport plays a minor role in soils compared to diffusive transport initiated by concentration differences. Diffusion itself is a passive process where molecules move randomly through space. In case of a concentration gradient this movement results in a vector directed from high to low concentration. This process is described by Fick in his first law:

$$J = D\frac{\delta c}{\delta x}$$ (2)

where J is the diffusion flux, D is the diffusion coefficient or diffusivity, c is the concentration of the gas and δx is the observed distance. D is proportional to the transfer velocity k_i of the diffusing particles, which depends on Temperature T and wind velocity:

$$J = k_i\frac{\delta c}{\delta x}$$ (3)

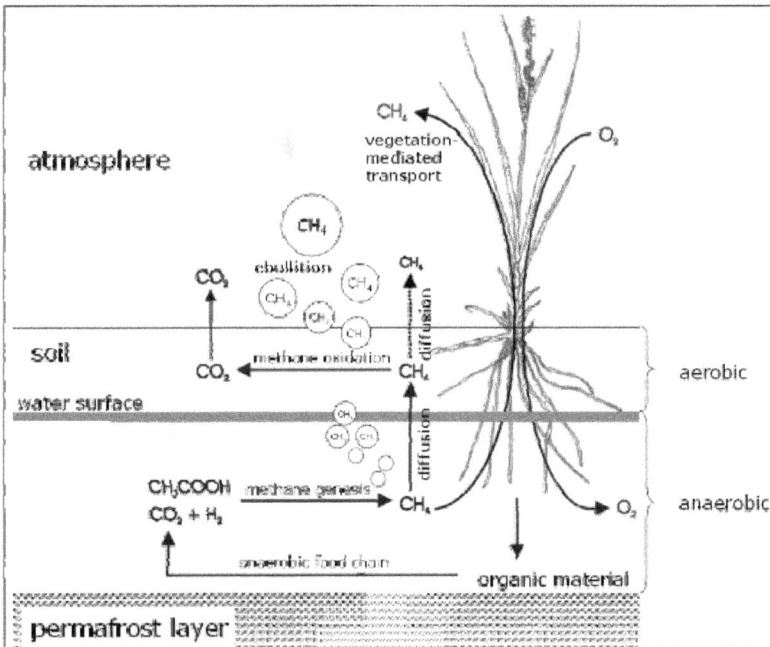

Figure 2-2: **Methane flux paths** [modified from *Kutzbach, 2000*]

Another emission pathway is vegetation-mediated transport (Figure 2-2). Aerenchyms of vascular plants ventilate oxygen from the atmosphere to the plants' roots situated in anaerobic layers. In the other direction aerenchyms enable a fast efflux of methane to the atmosphere, thus bypassing oxidation in aerobic soil horizons [*Kutzbach 2000*].

As solubility of methane in water is quite low, soil water can oversaturate. If the sum of partial pressures reaches a higher value than the one of hydrostatic pressure, bubbles grow and emit to the atmosphere. This study refers to this process as ebullition (Figure 2-2). *Scheffer and Schachtschabel [1998]* refer to it as free convection.

3   Material and methods

3.1   Investigation area

3.1.1   Lena River Delta

The investigation site is located in Arctic Siberia on Samoylov Island in the southern part of the Lena River Delta, Russia (72° 22' N, 126° 29' E). The Lena River is one of the longest rivers in the world and thus has of the largest watersheds. From its source at 1640 m above sea level (a.s.l.) in the Baikal Mountains south of the Central Siberian Plateau, 20 km west of Lake Baikal, the Lena flows northeast, being joined by the Kirenga River and the Vitim River. From Yakutsk it enters the lowlands, joined by the Olyokma River and flows north until joined by its right-hand affluent, the Aldan River. The Verkhoyansk Range deflects it to the north-west; then, after receiving its most important left-hand tributary, the Viljuj River, it makes its way nearly due north to the Laptev Sea in the Arctic Ocean. Altogether the total length of the Lena River is estimated at 4400 km. It drains an area of 2.49 million km² [*Alabyan et al., 1995*] and thus has an average annual discharge to the Laptev Sea of approximately 5.3×10^{11} m³ [*Peterson et al., 2002*]. Up to 35% of this discharge runs off during a few weeks of spring flooding in June [*Ivanov and Piskun, 1995*]. The Lena River Delta itself covers an area of 32000 km², making it the largest delta in the Arctic and one of the largest in the world [*Walker, 1998*]. It is considered to be a key of the area for carbon and climate dynamics of the Arctic because of its position at the interface between the Eurasian continent and the Arctic Ocean [*Rachold et al., 2000, 2002; Yang et al., 2002*]. Part of the area is protected in the Lena Delta Wildlife Reserve.

The delta area consists of more than 1500 islands and 800 river channels of various sizes (see Figure 3-1). Geomorphologically, it can be divided into three river terraces of different age and various floodplain levels [*Grigoriev, 1993; Schwamborn et al., 2002*]. Recent delta landscapes are represented by the youngest terrace of Late-Holocene age and the active floodplains [*Are and Reimnitz, 2000*], which can be found predominantly in the central and eastern part of the delta. They cover about 65% of the total delta area.

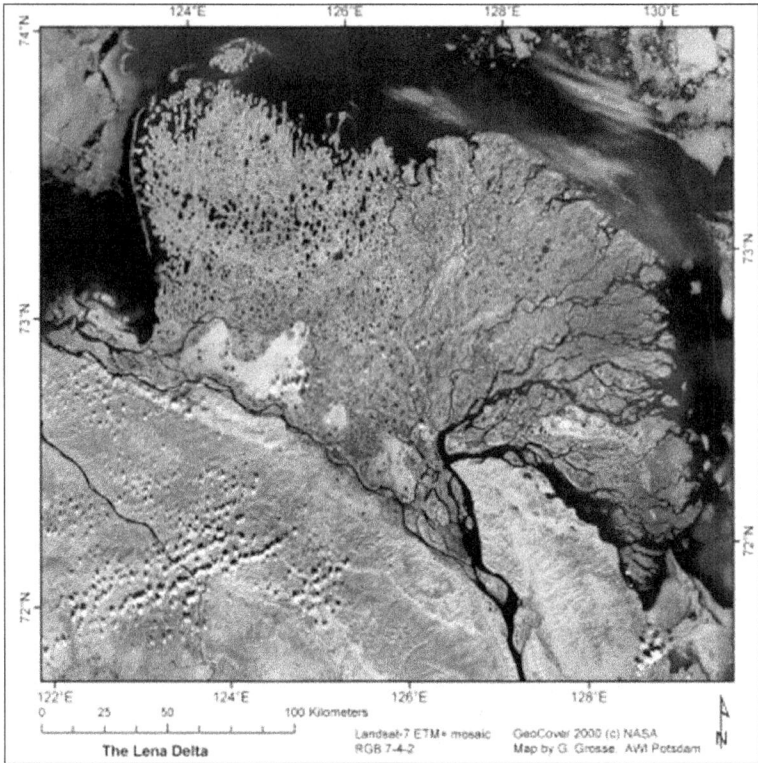

Figure 3-1: **Landsat-7 ETM+ mosaic of the Lena River Delta** - Samoylov Island is indicated by the red circle (Map by Guido Grosse, AWI Potsdam).

The second terrace can be found in the western part of the delta and is also known as Arga complex. The third terrace, consisting of late quaternary sands and its overlying ice complex, can mainly be found at the southern border of the delta [*Are and Reimnitz, 2000*].

3.1.2   Samoylov Island

Samoylov Island (72° 22′ N, 126° 29′ E) is situated at one of the main river channels, the Olyenyokskaya Channel, in the south-central part of the delta, where the bedrock of the North Siberian mainland is still in range of sight from the island. The permafrost in the area reaches depths of up to 600 m and is characterized by very cold temperatures between -13°C and -11°C [*Kotlyakov and Khromova, 2002*]. Samoylov Island is considered to be representative for the modern delta landscape. The triangular-shaped island has a size of 7.5 km² and can be divided into two main geomorphological units. The western part (3.4 km²) represents a recent floodplain annually flooded in spring by the river. Its elevation ranges from 1 to 5 m above sea level (a.s.l). The eastern part consists of the sediments of the Late-Holocene river terrace and represents a higher level than the younger floodplain with its elevation from 10 to 16 m a.s.l. (Figure 3-2), which is reached by the Lena River only during rare and extreme flooding events [*Kutzbach, 2006*]. The terrace is characterized by typical wet polygonal tundra. The relief appears to be plain on a mesoscale. But due to the development of ice wedge polygons, the surface is structured by a pronounced microrelief with elevation changes of up to 0.5 m within a few meters. Less frequently, slopes of 2-3 m within a few meters can be found at the shores of thermokarst lakes.

Figure 3-2: **The two main geomorphological units of Samoylov Island** (Photograph by Torsten Sachs, AWI Potsdam).

Figure 1-2 and Figure 3-2 show the net structure of wet polygonal tundra. The regular pattern of polygon centres and rims as well as the differing stages of development of the polygons has a strong influence on the water regime and other environmental conditions. For example, at depressed polygon centres, water is retained due to the underlying permafrost layer and surrounding ice wegdes, while elevated centres and polygons rims are much better drained and ventilated. Other parameters such as soil moisture, active layer depth, and soil texture, as well as the vegetation cover vary significantly within a few meters (Figure 3-3) Thus, every single polygon represents its own micro-site characteristics.

Figure 3-3:   **Typical profile of a polygon-micro-site** – (Photograph by AWI Potsdam)

3.1.3   Climate

The climate in the Lena River Delta is true-arctic, continental, and characterized by very low temperatures and low precipitation. The meteorological station Stolb, 20 km east of Samoylov Island, recorded a mean annual temperature of -12.3°C and a mean annual precipitation of 227 mm from June 2001 to May 2004. In this period the average temperatures of the coldest month (Feburary) and the warmest month (July) were -34.17°C and +11.71°C (Figure 3-4-A) [*HMCR, 2007*], showing the large amplitude between the seasons that is typical for continental climates. Due to its position above the Polar Circle, the sun does not set in summer. This phenomenon is referred to as Polar Day. Vice versa the time period of winter solstice when sun does not rise above the horizon at extreme latitudes on the northern hemisphere is referred to as Polar Night. Polar Day on Samoylov Island begins at May 7th and ends at August 7th, Polar Night lasts 74 days from November 15th to January 28th. More than 50% (52.1%) of the precipitation occurs in the period of active photosynthesis from May to September. The rest falls as snow and does not run off until next spring. However, snow drift due to frequent and strong winds is common. Another important process in the Lena River Delta is snow sublimation [*Boike et al., 2003*].

Comparing data from the meteorological station of Stolb to the one at Tiksi, 110 km south-east of Stolb, shows that temperatures at both stations were very similar during the period from June 2001 to May 2004. As there are no long-term data available for the meteorological station Stolb, one can assume that long-term records for Tiksi can be transferred to the soutern central delta area. A comparison between the long-term record and the younger period shows, that the region was significantly warmer during the years 2001...2004 (Figure 3-4-B,C). This can be n as an indicator for the warming of this region. This development is further predicted by global circulation models, which calculate an increase of mean annual temperatures of about 5 °C in the Lena River Delta until the end of the 21st century [*ACIA, 2004*].

Although precipitation is low, the climate is classified as humid due to low evaporation caused by low ambient temperatures. The synoptic weather conditions are characterized by the delta's position at the border between the Arctic Ocean and the Siberian mainland. In winter the Lena River Delta region is influenced by the Siberian High, a cold anticyclone usually formed over eastern Siberia. The Siberian High as main cause for extremely low temperatures in Yakutia, can even influence temperatures in central Europe in some winters. During summer the high is replaced by a strong low over central Siberia which does not completely reach up to the Lena Delta due to low pressure systems over the Arctic Ocean [*Kutzbach, 2006*].

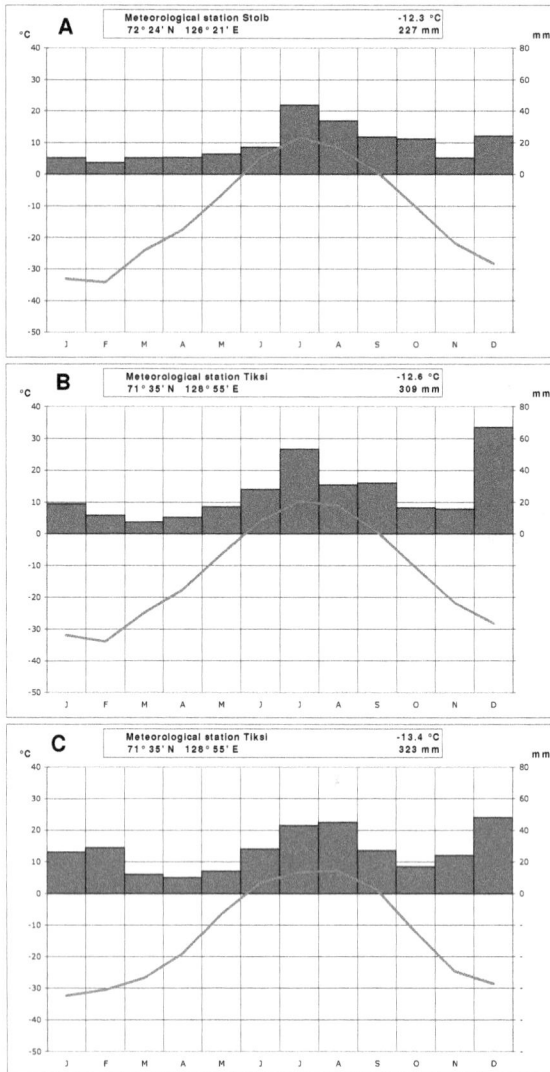

Figure 3-4: **Climate charts for the meteorological stations Stolb and Tiksi** – A, B monthly averages of the period June 2001 to May 2004 for meteorological stations Stolb and Tiksi [*HMCR, 2007*], C long-term monthly averages of the 30 year-period 1961...1990 for Tiksi [*ROSHYDROMET, 2007*].

3.2 Investigation site

The closed chamber measurement campaign was conducted in immediate proximity to the eddy covariance system in the centre of Samoylov Island. To avoid disturbances of the sensible polygon environment and vegetation cover, wooden boardwalks were installed in previous seasons to avoid flux discontinuities caused by the presence of the observant. The boardwalks are clearly visible in Figure 3-5. Altogether, 15 plots in four different polygon centres and on one polygon rim (each site with three plots) were investigated in almost daily iteration. Water-filled cracks in between single polygons were measured twice at the end of August to estimate the potential for methane ebullition from these thermokarst features (Figure 3-6).

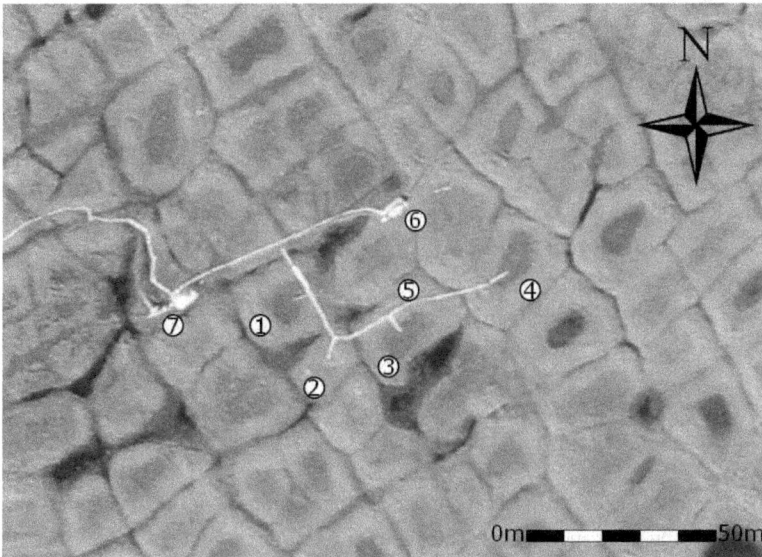

Figure 3-5: **Aerial view of investigation site from approx. 500 m a.s.l. -** ① low centre polygon P I, ② high centre polygon P II, ③ low centre polygon P III, ④ low centre polygon P IV, ⑤ polygon rim P V, ⑥ eddy covariance system installation, ⑦ tent for technical equipment (Photograph by Julia Boike, AWI Potsdam).

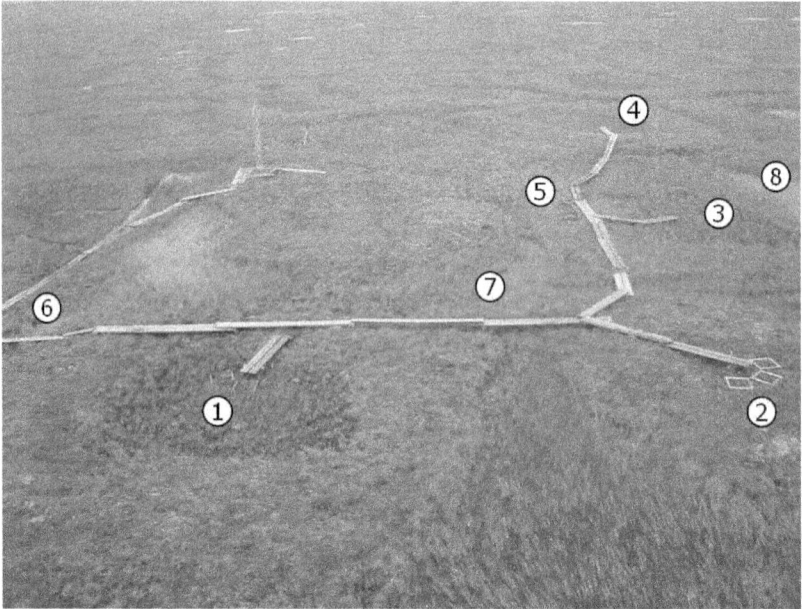

Figure 3-6: **Investigation site from south-western approach** - ① micro-site P I, ② micro-site P II, ③ micro-site P III, ④ micro-site P IV, ⑤ micro-site P V, ⑥ micro-site W I, ⑦ micro-site W II, ⑧ micro-site W III (Photograph by Torsten Sachs, AWI Potsdam).

3.2.1 Polygon micro-sites

The five micro-sites do not vary much in their composition of vegetation with regard to the plant morphologic categories shrubs, herbs, mosses and lichens. All of them are mainly covered by mosses. Nevertheless they do vary in plant species coverage due to their different genesis and soil-hydrological conditions. A detailed list of the identified plant species can be found in the Appendix (Table 7-1).

3.2.1.1 Micro-site P I

Micro-site P I is a low centre polygon in which the boundary of polygon centre and rim is clearly visible by its different heights, although the north-eastern part of the rim appeared to be in an early stage of degradation. As all other polygons, the polygon is completely covered with mosses (Table 3-1 and Figure 3-7). A vegetation survey conducted on July 5th shows that micro-site P I is mainly covered by one species, *Drepanocladus revolvens*, a moss species of the *Amblystegiaceae* – family. This species is characterized by its long, tapered leaves, which are strongly curled to the extent of almost forming a circle. The leaves are orientated to one side of the stem. It is a robust plant and is usually tinged with orange, crimson or purple, mixed with yellowish-green. Among *Drepanocladus revolvens* other moss species were surveyed in much smaller numbers: *Meesia triquetra* (threeranked humpmoss), *Calliergon giganteum* (arctic moss) and *Aulacomnium turgidum*. In the herb category, three species were found in this first polygon. In total, the surface of micro-site P I is covered with 10% herbaceous plants, predominantly with *Carex chordorrhiza* or creeping sedge. *Carex chordorrhiza*, as a member of the sedge family (*Cyperaceae*), is a perennial, herbaceaous, grass-like plant of sphagnum bogs and fens. It is named for its unusual prostrate, creeping growth pattern, where decumbent stems from the previous year send up new shoots from the nodes. Often these horizontal, older shoots are overgrown by moss and are just below the surface, appearing to be rhizomes [*Massachusetts Division of Fisheries & Wildlife, 2004*]. Among this dominant species *Carex concolor* and *Comarum palustre* (marsh cinquefoil or purple marshrocks) were also surveyed.

Table 3-1: **Cover of plant morphological categories on micro-site P I.**

	shrubs	herbs	mosses	lichens
Micro-site P I		10%	100%	

Figure 3-7: **Vegetation in plots of micro-site P I.**

The water level at this micro-site was at or above the soil surface throughout the measurement period, so that soil moisture is considered to be 100%. As in all other low centre polygons, the water level showed fast reaction to precipitation events. Changes in water level were included in the chamber headspace volume estimation.

3.2.1.2   Micro-site P II

As a characteristic high centre polygon micro-site P II is covered with completely different vegetation than the others. Though the plants are still unmistakeable indices for bog and fen landscapes as they are typical for the arctic tundra and though this polygon is completely covered with mosses as the other surveyed polygons, the predominant moss and herb species found at micro-site P II are different. Micro-site P II was the only polygon covered with single woody shrubs (see Table 3-2 and Figure 3-8), namely *Salix glauca*. *Salix glauca* (glaucous willow) as a member of the willow family (*Salicaceae*) can reach up to 150 cm in height, but here it just crouches on the ground. Twigs of the season usually have a dark or reddish colour underneath the pubescence. Its leaf blades are elliptic with a blue-white and waxy coating below [*Hitchcock et al., 1969*]. Nevertheless, the moss category is predominant in polygon P II. Here, *Hylocomium splendens* was identified as the predominant species with about 85% coverage. *Hylocomium splendens* is a widely-spread moss which can be found on almost every continent, and even in New Zealand [*Frahm et al., 2004*]. *Tomentypnum nitens* and *Campylium stellatum* from the class of byropsidae cover the gaps in this "moss carpet" on polygon P II. Herbs appear in various but very small numbers. Even rarer is the appearance of one single lichen species at this micro-site.

Table 3-2:   Cover of plant morphological categories on micro-site P II.

	shrubs	herbs	mosses	lichens
Micro-site P II	1%	2%	100%	1%

Figure 3-8:   Vegetation in plots of micro-site P II.

Due to the high centre character of this micro-site, the polygon is much better drained and usually does not have any standing water.

3.2.1.3 Micro-site P III

The vegetation coverage of Polygon P III is very similar to Polygon P I (Figure 3-10). *Drepanocladus revolvens* is the predominant moss on this micro-site followed *Meesia triquetra* (threeranked humpmoss), *Calliergon giganteum* (arctic moss) and *Aulacomnium turgidum*. The main difference of P III is the larger number of herbs. Though the same three species *Carex chordorrhiza*, *Carex concolor* and *Comarum palustre* appear, especially *Carex concolor* grows more successfully on polygon P III. Thus, micro-site P III can be described as the most herbaceous polygon of the survey (Table 3-3).

Table 3-3: **Cover of plant morphological categories on micro-site P III.**

	shrubs	herbs	mosses	lichens
Micro-site P III		20%	100%	

Figure 3-9: **Vegetation in plots of micro-site P III.**

The water level in micro-site P III remained at or above soil surface level similar to micro-site P I. Soil moisture is therefore considered to be 100%. The ice wedge of the south-eastern rim was completely degraded, resulting in a deep thermokarst crack with open water and a hydraulic connection of the polygon centre to the surrounding cracks and troughs.

3.2.1.4 Micro-site P IV

Polygon P IV was similar to P I and P III concerning its low centre character and the water saturated soil (Figure 3-10). Differences could be observed in the vegetation coverage. Though P IV is entirely covered by mosses as well (see Table 3-4), the predominant species is one that only appears in this polygon: *Scorpidium scorpidiodes* (scorpion moss) from the class of Byropsidae covers 100% of the polygon surface and –neglecting a very small amount of *Callergion giganteum*- is the only moss growing in this low centre polygon. Among this unusual moss the same three herbs as in polygon P I and polygon P III appear in similar amounts: *Carex chordorrhiza*, *Carex concolor* and *Comarum palustre*.

Table 3-4: **Cover of plant morphological categories on micro-site P I V.**

	shrubs	herbs	mosses	lichens
Micro-site P IV		15%	100%	

Figure 3-10: **Vegetation in plots of micro-site P I V.**

Micro-site P IV is different with regard to the water level as well. It constantly had the highest water level, due to the fact that its rim did not show any signs of degradation or hydraulic connection to surrounding troughs. Especially at the beginning of the measurement period, P IV could have been characterized as a polygon pond, as its water level was never below 10 cm above ground (cmp. Figure 4-8).

3.2.1.5 Micro-site P V / Polygon Rim

As the only polygon rim in the study micro-site P V does not differ as much as one may expect from the other surveyed micro-sites (Figure 3-11). Even the same predominant moss species *Hylocomium splendens* can be found here. Another dominant moss species is *Rhytidium rugosum* ("rabbit paw moss"), which forms yellow-green to golden-brown-green mats with branched stems 6-10 cm long. Its leaves are 3-4 mm long [*Schofield, 1992*]. A large variety of different herbaceous species was identified on this rim. In addition, several types of lichens were found at this micro-site only.

Table 3-5: **Cover of plant morphological categories on micro-site P V.**

	shrubs	herbs	mosses	lichens
Micro-site P V		7%	100%	2%

Figure 3-11:   Vegetation in plots of micro-site P V.

Due to rapid drainage of precipitation into the surrounding polygons and troughs, the rim was the "driest" micro-site.

3.2.2   Crack/ trough micro-sites

In a former study CH_4 emissions from thermokarst lakes on Samoylov Island were observed [*Spott, 2003*]. Water-filled cracks and troughs can represent early stages of these lakes and are common, especially in the eddy covariance footprint. *Spott* [*2003*] differentiated plant-mediated transport, ebullition and diffusive transport in his study.

In order to get a rough estimate of methane ebullition or potential methane ebullition from water-filled polygon troughs, occasional measurements were conducted on troughs in the area using a special floating chamber (Figure 3-12). Three troughs, which were situated between the single polygons, were chosen for this observation (Figure 3-6). Measurements took place on August 26th 2006 and during continuous 15...24 hour night-time measurements from August 26th 2006...September 2nd 2006. If possible - depending on size and accessibility - these three troughs were subdivided into areas of non-vegetated/vegetated water surface and vegetated shore-zones. The six micro-sites on polygon troughs were:

Micro-site W I: a small crack merely of the size of the used floating chamber, massive submerged vegetation, partially emerging vegetation

Micro-site W II a: shore zone of the second larger trough, massive submerged vegetation, partially emerging vegetation

Micro-site W II b: in about 1m distance from the trough's edge, still massive submerged vegetation, non-vegetated water surface

Micro-site W III a: shore zone of a larger thermokarst feature, which is situated immediately next to polygon micro-site P III, less submerged vegetation than the comparable micro-sites W I and W II a, partially emerging vegetation

Micro-site W III b: in about 1m distance from the trough's edge, less submerged vegetation, non-vegetated water surface

Micro-site W III c: in about 2m distance from the trough's edge, nearly no submerged vegetation, non-vegetated water surface

Figure 3-12: **Measurement at thermokarst crack micro-site W III** – setup consists of ① chamber unit, ② pump unit, ③ sensor unit INNOVA 1412, contained in the visible box (Photograph by Torsten Sachs, AWI Potsdam).

These measurements were added to the regular measurement program as an ad hoc preliminary survey in response to a Nature article by *Walter et al* [*2006*] describing the importance of methane emission by ebullition from Siberian thaw lakes. Due to their preliminary nature, these measurements had a lower priority in this study. Every long-term measurement was completed with a simple test for potential ebullition by disturbing the bottom of the cracks/troughs using a steel probe and thus forcing the release of methane stored in the sediment.

3.3   Sampling procedure

The closed chamber method was chosen for flux measurements in the vegetated and wet peatland of Samoylov Island. At each micro-site three PVC chamber collars enclosing an area of 0.25 m² had been inserted 10-15 cm into the active layer in 2005. The chambers were placed in a water-filled channel on top of each collar, which acted as a seal and ensured a closed system during the measurement. Chambers were made of opaque PVC and clear PVC, respectively, for dark and light measurements. Chamber volume was 12.5 l at the high centre and rim micro-sites and 37.5 l at the other sites, where higher vegetation did not allow for the use of small chambers. According to *Matson & Harriss* [*1995*] a closed chamber system consists of three major parts, i.e. the chamber itself, a sensor unit (see Chapter3.4) which analyzes the sample air, and a climate control system. The latter, however, has not been sufficiently implemented in AWI chambers. Instead, a 12 V pump maintained air circulation within the system and a temperature logger recorded temperature inside the chamber at high temporal resolution (cmp. Chapter 3.6.1.5). In order to prevent adverse chamber effects on humidity and temperature while still generating enough data points for a meaningful analysis, chamber closure time was limited to 6-8 minutes. Figure 3-13 shows the simple setup of the closed chamber system. The sensor unit draws sample air through the water filter into its measurement chamber. The sampled air is then released back into the closed chamber. Along with the air ventilation this setup provides a closed cycle in which gas concentration change can occur only between the soil-atmosphere-interface.

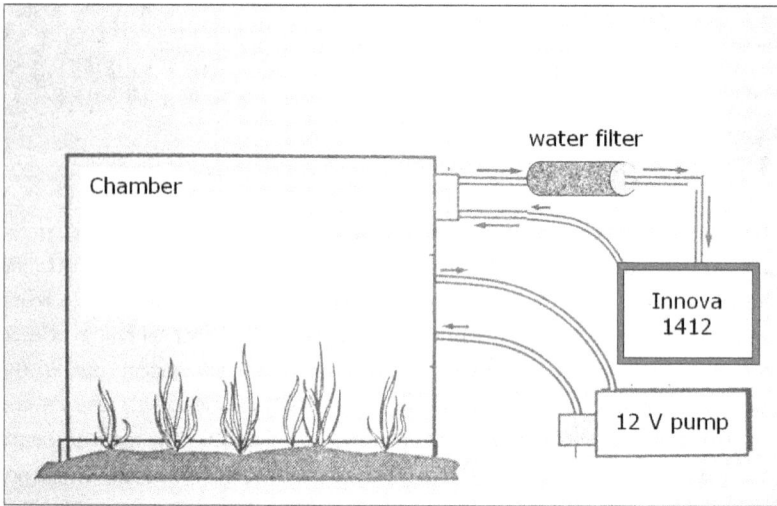

Figure 3-13: Sketch of sampling setup [modified from *Matson & Harris, 1995*]

3.4 Sensor unit

Chamber headspace air was analyzed by photoacoustic infrared spectrometry using an INNOVA Field Gas Monitor 1412 (LumaSense Technologies A/S - INNOVA Airtech Instruments, Ballerup, Denmark). The technical principle of the photoacoustic spectrometer is to draw sample air into a measurement chamber, where infrared (IR) radiation from an IR-source passes through a chopper and an optical filter into the chamber. The IR radiation is absorbed by gas molecules and generates specific heat and pressure variations. These pressure variations correspond to the choppers frequency creating a pressure wave, which then can be detected by highly sensitive microphones (see Figure 3-14). The microphone signal is then post-processed and the measurement result can be calculated [*INNOVA Airtech Instruments, 2006*]

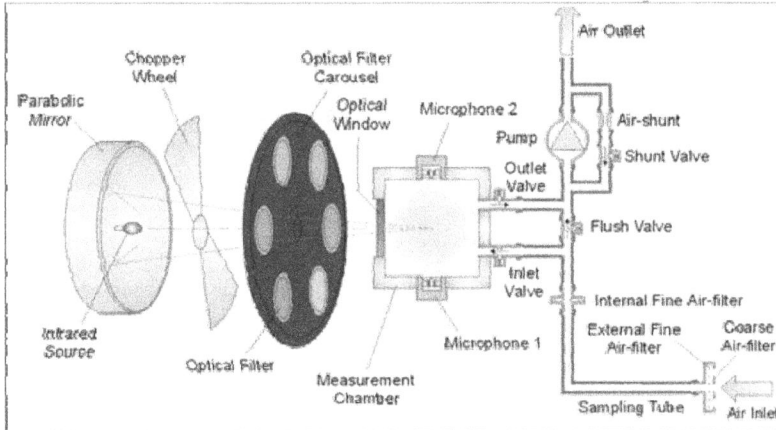

Figure 3-14: **Technical principle of photoacoustic spectroscopy** [*INNOVA Airtech Instruments, 2006*]

The following configurations were set at the INNOVA 1412:

- Auto Flush on,

- water compensation on,

- tube length 20.0 m.

In addition to methane, air pressure and water vapour were measured simultaneously. Due to an existing cross interference of water vapour and methane, the sample air was dried before reaching the sensor unit using a filter installed between chamber and sensor unit. The filter was filled with 0.3 nm molecular sieve (beads with moisture indicator; Merck KGaA, Darmstadt, Germany).

3.5 Measurement campaign

As carbon dioxide was measured simultaneously, all measurements were conducted with clear plexiglas chambers and then repeated with opaque chambers, in order to be able and separate photosynthesis and respiration. For methane, which is reported in this study, no difference was observed between fluxes measured with clear and opaque chamber, resulting in two flux measurements per plot.

The closed chamber measurement campaign of this study was conducted from July 12th 2006 until September 19th 2006. During this period there were 47 days of observation. On average, one chamber headspace sample per 45-60 seconds was analyzed during the 6-8 minute chamber closure, resulting in 8-10 concentration data points per plot. In total, this amounts to over 10,000 samples (Table 3-6). With three plots on each micro-site and both clear chamber and opaque chamber measurements on each plot, this results in 30 chamber measurements of approximately 10 minutes for the entire site to be covered. Therefore, every plot could only be measured once in numerical order (plot 1 clear chamber, plot 1 opaque chamber, plot 2 clear...plot 15 dark). For reasons of better comparison the order of measurement was never changed.

Table 3-6:   Campaign statistics

Observation beginning	12/07/2006
Observation ending	19/09/2006
Days	69
Total number of days in observation	47
Days in observation (%)	68
Total sampling time (hours)	256.47
Total number of samples	19705
averaged observing time/day (hours)	5.46
Total sampling time in water-filled cracks (hours)	127.73
Total number of samples in water-filled cracks	8880
Sampling time in water-filled cracks (%)	49.80
Samples in water-filled cracks (%)	45.06
Total sampling time in plots (hours)	128.75
Total number of samples in plots	10825
Sampling time in plots (%)	50.20
Samples in plots (%)	54.94

3.6   Environmental parameters

3.6.1   Meteorology

General meteorological parameters were logged continuously throughout the campaign in high temporal resolution at the automated eddy covariance system close to the investigation site and a long-term climate monitoring station approximately 500 m south of the site.

3.6.1.1   Horizontal wind speed *u* and friction velocity *u*∗

Wind velocity components were measured using a three-dimensional sonic anemometer Solent R3 (Gill Instruments Ltd., Lymington, UK). The instrument was installed 4 m above ground level. As horizontal wind speed u and the directly correlated friction velocity u_*, respectively, were found to be important controls on ecosystem scale methane emission [*Sachs et al., 2007*], these parameters were also of interest for this study. One hour averages of friction velocity (a measure for near-surface turbulence) were used in the analysis.

3.6.1.2   Barometric pressure *p*

The sensor for barometric pressure was a RPT410 (Druck Messtechnik GmbH, Dresden, Germany). The data set forming the base of in this study used barometric pressure p was recorded in 1 hour intervals.

3.6.1.3   Air temperature *T*air

The MP103A (ROTRONIC AG, Bassersdorf, Switzerland) was used for measuring air temperature. One hour averages were used for the analyses in this study.

3.6.1.4   Photosynthetically Active Radiation *PAR*

Photosynthetically Active Radiation (PAR) was logged hourly throughout the season and additionally in 60 second intervals from July 24th to September 19th using a QS2 (Delta-T Devices Ltd., Cambridge, UK). *PAR* designates the spectral range of solar light from 380 to 780 nm that is useful to terrestrial plants in the process of photosynthesis [*Gates, 1980*].

3.6.1.5   Chamber temperature T_{ch}

Temperatures within the chamber deployment were logged with a Minidan Temp0.1 (ESYS GmbH, Berlin, Germany) in 5 second intervals. This parameter chamber temperature T_{ch} was one criterion in the ensuing flux calculation, where then one averaged value per deployment was used.

3.6.2   Soil climate

Water level /soil moisture, soil temperatures in various depths and thaw depth (active layer thickness) were recorded daily at each micro-site.

3.6.2.1   Water budget

Precipitation *PR* was measured at the long-term climate station approx. 500 m south of the investigation site using a tipping bucket rain gauge model no. 52203 (R. M. Young Company, Traverse City, USA). Cumulative daily precipitation was used for analyses in this study.

Depending on whether a micro-site was inundated or well-drained either the water level *W* or the soil moisture θ of each polygon was recorded. Water levels were measured using a common electrical plummet< which was lowered into simple observation wells installed in the active layer. Positive values indicate water levels above the soil surface, while negative values indicated water levels below the soil surface. Soil moisture was measured using a Theta Probe type ML2x (Delta-T Devices Ltd., Cambridge, UK). Due to possible within-site spatial variability, an average value of three iterations at different spots was recorded.

The height of the water column W_{col} was calculated from active layer thickness and water level and introduced as an additional parameter for two reasons: first, methane production takes place under anaerobic conditions (e.g. water saturation). Second, standing water above the soil surface can hamper methane emission due to the low diffusive fluxes, unless enough emerged vegetation is present to act as a conduit. Therefore, the height of the actual water column is of interest for further flux analysis.

3.6.2.2   Soil temperature $T_{soil\ i}$

Soil temperatures $T_{soil\ i}$ were recorded in 5 cm depth intervals starting directly underneath the surface and continuing all the way to the top of the permafrost table. Living moss layers were defined to be above the soil surface. Measurements were conducted with the soil thermometer DTTF (Umwelt- und Ingenieurtechnik GmbH, Dresden, Germany).

3.6.2.3   Active layer thickness *L*

Active layer thickness was measured by driving a metal rod into the soils of the micro-sites until it hit the permafrost table. The average value of three iterations at different spots was recorded to lessen the likelihood of recording an extreme alteration of the potentially uneven permafrost. The metal rod was marked with a centimetre scale for easy and direct reading of the thaw depth.

3.7   Data processing and quality control

Methane fluxes were calculated using a MatLab®-script developed at the University of Greifswald by *Lars Kutzbach et al.* [2007]. There is still discussion about how to calculate fluxes from the concentration change under closed chambers. Many studies use simple linear regression [*Vourlites et al., 1993; Oechel et al., 1993, 1998, 2000; Jensen et al., 1996; Alm et al., 1997, 2007; Goulden and Crill, 1997; Christensen et al., 1998; Tuittila et al., 1999; Maljanen et al., 2001; Xu and Qi, 2001; Bubier et al., 2002; Pumpanen et al., 2003; Burrows et al., 2004; Heijmans et al., 2004; Reth et al., 2005; Laine et al., 2006; Wang et al., 2006*]. However, evidence is quickly growing, that linear regression is not appropriate due to the inherent disturbances of natural fluxes caused by deploying closed chambers. The MatLab®-script developed by *Kutzbach et al., [2007]* accounts for some of these disturbances and applies a set of different linear and non-linear models for flux calculation. Thorough analysis of residuals showed that linear regression was frequently not appropriate for the determination of carbon dioxide fluxes by closed-chamber methods and found an exponential regression model to be better suited. Since many of the underlying disturbances and limitations of the method are not likely to be limited to specific gases, the exponential regression model developed by *Kutzbach et al.* [2007] was also applied for flux calculation in this study.

Where concentration data were too noisy to fit any regression model, measurements were discarded during flux calculation. Calculated fluxes were then thoroughly screened and all fluxes with a residual standard deviation greater than 0.3 ppm were excluded from further analysis. The remaining fluxes were then summarized for each micro-site by averaging the six daily measurements.

3.8   Model Development

Correlation and multiple linear regression analyses were used to identify environmental controls on methane fluxes at each micro-site. Due to some parameters being directly correlated to each other (such as soil temperatures in different depth), parameters were divided into six main categories: air temperatures (chamber and air temperatures), soil temperatures (soil temperatures of every layer), water and active layer (water levels / soil moisture, water column, and active layer), PAR, atmospheric parameters (friction velocity and pressure). Out of each category, the parameter which was most correlated with the methane flux was then selected for multiple linear regression analysis, described by the following equation:

$$F_{\text{microsite}}(t) \;=\; \beta_0 + \beta_1 x_1(t) + \dots + \beta_i x_i(t) \tag{4}$$

where $\beta_1 \dots \beta_i$ are coefficients for the category parameters $x_1 \dots x_i$.

In an iterative procedure, the least significant parameter in the multiple linear regression was dropped before repeating the regression with the remaining parameters. This stepwise elimination of non-significant parameters was done until only those parameters were left, that were significant at the 95% confidence level.

4   Results

4.1   Environmental conditions

4.1.1   Meteorology

4.1.1.1   Wind

The average friction velocity u_* was 0.41 m s^{-1} during the measurement period from July 12th ... September 19th. Maximum wind speeds of up to 24.49 m s^{-1} and associated high turbulence were reached during two distinct bad weather periods at the beginning of August and the beginning of September. (Figure 4-1).

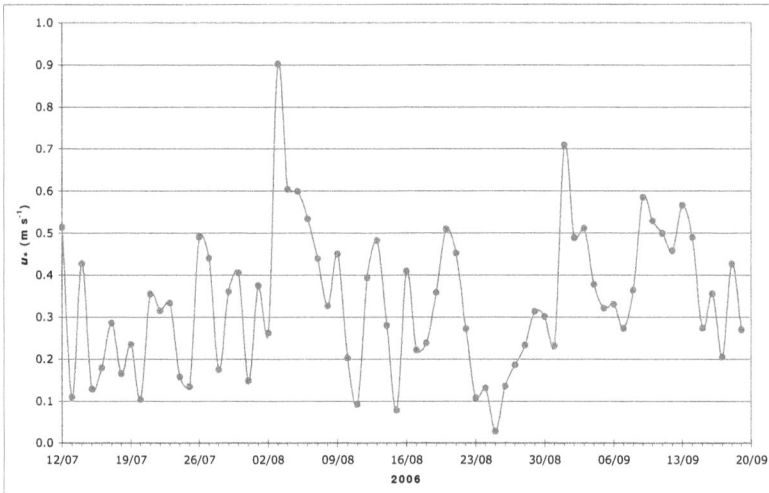

Figure 4-1:   **Daily averages of friction velocity u_* at the eddy covariance system.**

4.1.1.2 Barometric pressure

Average barometric pressure p was 100.54 kPa during the period from July 12th ... September 19th. The maximum pressure of 102.21 kPa was reached during the night from August 27th ... August 28th. The lowest pressure of 98.00 kPa was recorded on September 13th (Figure 4-2). In general, barometric pressure was rather low during the first half of September which was coincident with the bad weather conditions during this time.

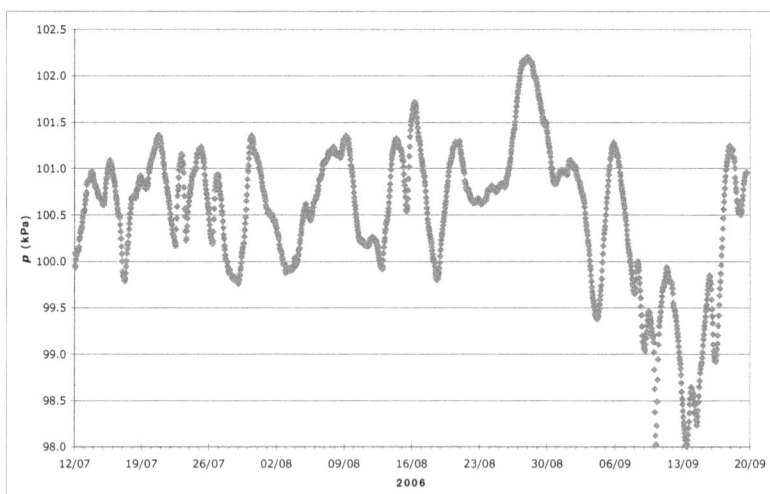

Figure 4-2: **Barometric pressure on Samoylov Island** - barometric pressure p recorded in 1 hour intervals during the survey period.

4.1.1.3 Air temperature

Average air temperature T_{air} was 5.3°C. Thus air temperatures were rather low compared to the data from 2001...2004 of the same time period, when air temperatures averaged 7.6°C. The maximum of 26.0°C was reached at noon of July 31st. The lowest temperatures were recorded at the early hours of the morning of September 9th at -7.2°C (see Figure 4-3).

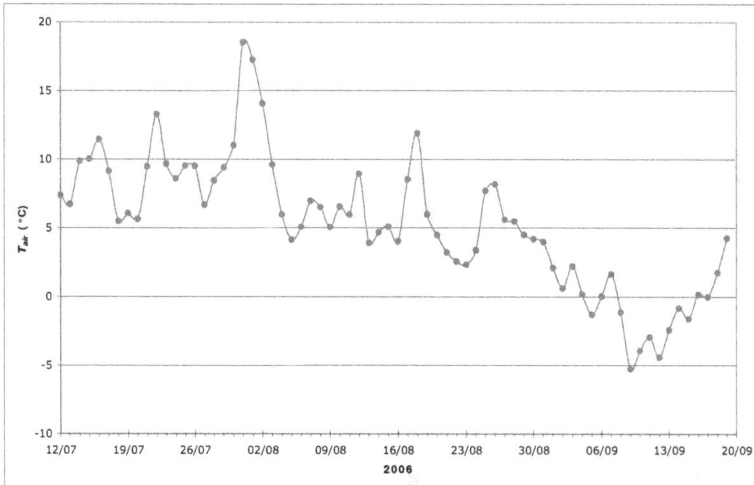

Figure 4-3: Daily average air temperatures T_{air} at the eddy covariance system in 2 m above surface.

4.1.1.4   PAR

Cloudy conditions are very common on Samoylov Island, which is evident from the PAR record. Sunny days without cloud cover appear as smooth sinus curves with high amplitudes in Figure 4-4. It shows that July 13th, 24th, 25th, August 1st, 17th, 18th, September 15th, 18th, and 19th were the only dates with a clear blue sky during the campaign. In total these are just 9 days in a measurement period of more than 2 ½ months.

Figure 4-4 also shows the effect of the Polar Day: during nights in July, PAR rarely reaches zero and is still measurable at midnight. When the sun sets again in August, PAR reaches zero and shows the usual alteration of day and night.

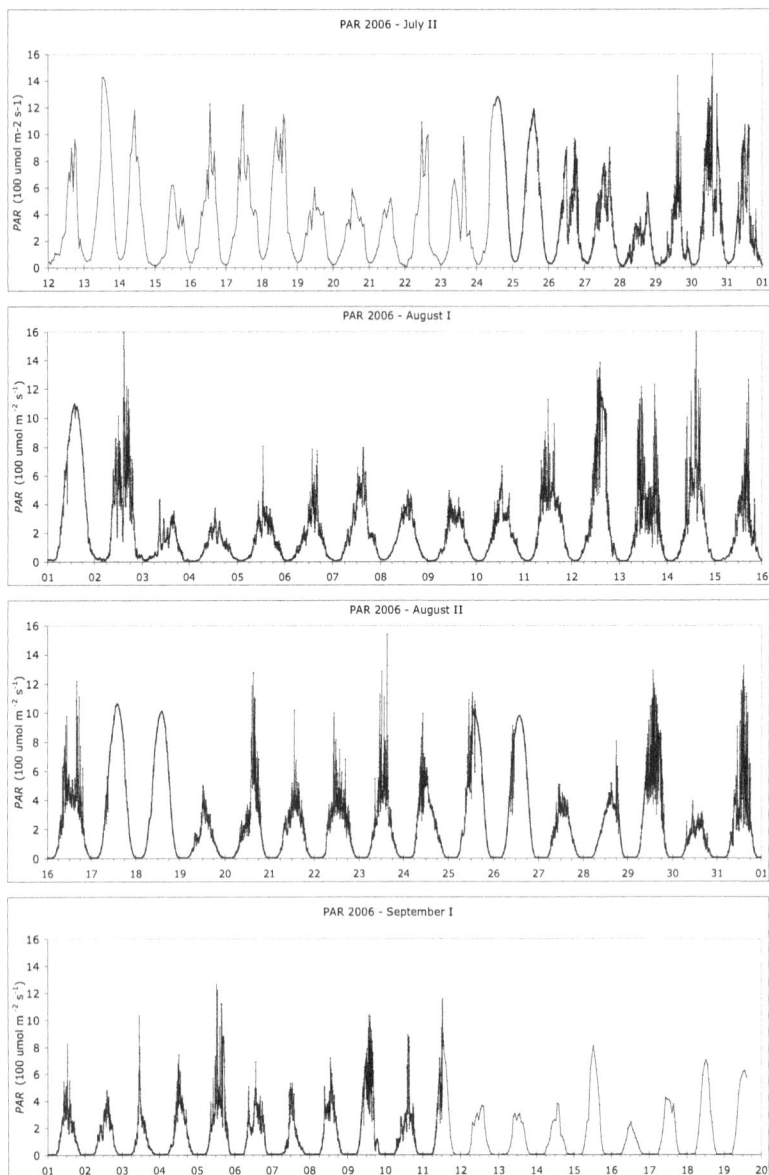

Figure 4-4 :   PAR recording summer 2006 on Samoylov Island.

4.1.1.5 Chamber temperature

Though chamber deployment periods were kept as short as possible, temperatures underneath the chamber increased rapidly during could-free conditions. Especially when using the clear chamber under suny conditions a strong temperature gradient was observed. Figure 4-5 shows the steep slope of chamber temperatures T_{ch} on two days with clear sky conditions (July 24th and August 1st), compared to the chamber temperature record on a day with more cloudy conditions (August 3rd) (cmp. Figure 4-4). The graph shows averaged temperatures for each micro-site, but differentiates between clear (*cl*) and opaque (*op*) chambers. The maximum gradient is reached during the clear chamber measurements at micro-site P III on July 24th. Here, chamber temperature increased by nearly 10°C (9.1°C) during a chamber deployment period of less than 8 minutes. But even when deploying opaque chambers the gradient is steep: during opaque chamber measurements at micro-site P III on July 24th temperature under the deployed chamber still increases by 5.3°C. On August 3rd, as an example for cloudy days, alteration of temperatures during chamber deployment is rather low. All graphs of T_{Ch} at the different micro-sites show a minor slope. The maximum gradient is found again during clear chamber measurements at micro-site P III, but here the increase amounts to just 1.6°C during a chamber deployment period of about 8 minutes. The increase under opaque chambers is 0.9°C at the same site. In addition, the starting chamber temperatures are not altered as significantly as they already are on July 24th and August 1st. Thus, here, the deployed chamber does not create such an artificial environment, as it has strong influence on the temperature under deployment on sunshiny days.

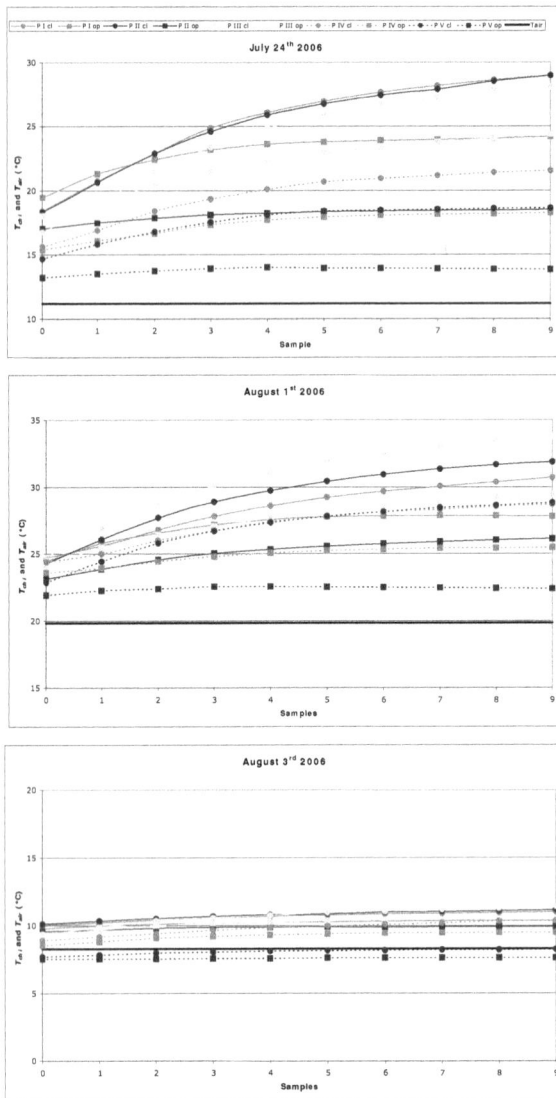

Figure 4-5: Chamber temperatures T_{ch} at micro-sites P I to P V – black horizontal line indicates average air temperature T_{air}, coloured lines indicate different micro-sites differentiating between clear chambers (dots) and opaque chambers (squares).

4.1.2 Soil climate

4.1.2.1 Water budget

Total precipitation during the measurement period was 130.48 mm. Compared to the available data from previous seasons (Chapter 3.1.3), liquid precipitation was above average, and most of it occurred during two main events (Figure 4-6). At the beginning of August, three consecutive days of rainfall yielded up to 23 mm a day causing water levels in the micro-sites to rise immediately (Figure 4-8). The second distinct event at the beginning of September brought a mixture of rain and wet snow when a storm system passed Samoylov Island.

The three low centre polygons P I, P III and P IV were always water saturated, while micro-site P II as a high centre polygon and micro-site P V as a polygon rim were rather well-drained. Standing water in the wells could only be found after precipitation events. Therefore, soil moisture was recorded at these sites and the water levels of P II and P V were excluded in Figure 4-7.

Snow started to accumulate on September 12th and reached depths of 8-10 cm in polygon centres and 2-6 cm on elevated areas. Due to advection of warmer air from the south, this early snow had disappeared again on September 18th.

4.1.2.2 Soil temperatures

Soil temperatures are directly dependent on air temperature but react with an increasing delay at greater depths. Comparing the graphs of soil temperature over time at the different depths also shows a decrease in amplitudes with increasing depth (Appendix Figure 7-1 and Figure 7-2). Figure 4-9 shows the temperatures for each of the 5 surveyed polygons at a level directly underneath the soil surface. All other soil temperature data can be found in the Appendix.

4.1.2.3 Active layer

Thaw depth increased continuously throughout July and August. When air temperatures fell below 0°C at the beginning of September (cmp. Figure 4-3), the active layer did not react immediately as can be seen in Figure 4-10. However, the cold period in the second week of September initiated refreezing from the bottom of the active layer in addition to freezing of the soil surface. In general, the three micro-sites with low-centre polygon character P I, P III and P IV were thawing about 80…90% deeper than the high-centre micro-site P II and the polygon rim P V due to the warm water bodies, differences in the vegetation cover and its insulation effect, and the presence or absence of ice wedges.

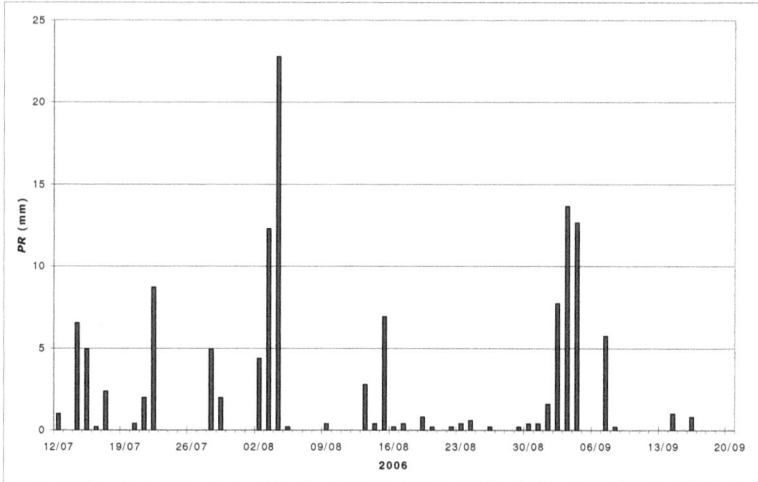

Figure 4-6: **Precipitation *PR* on Samoylov Island** – [*unpublished data, Boike, 2006*].

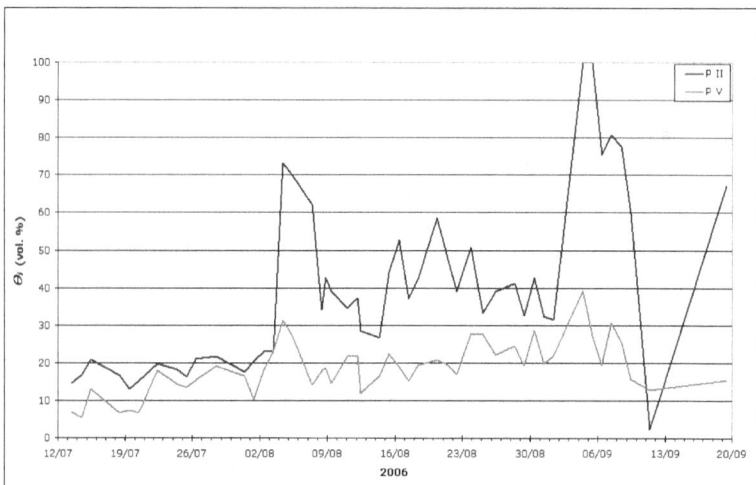

Figure 4-7: **Soil moisture *θ$_i$* of micro-sites P II and P V.**

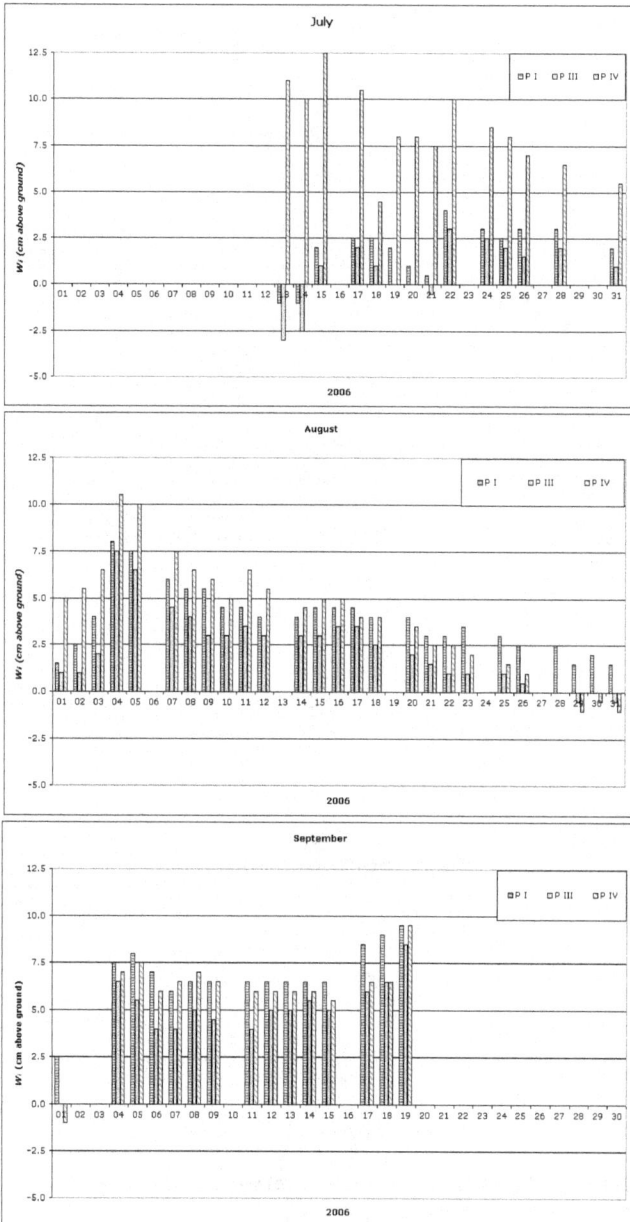

Figure 4-8 : Water levels W_i of micro-sites P I, P III and P IV.

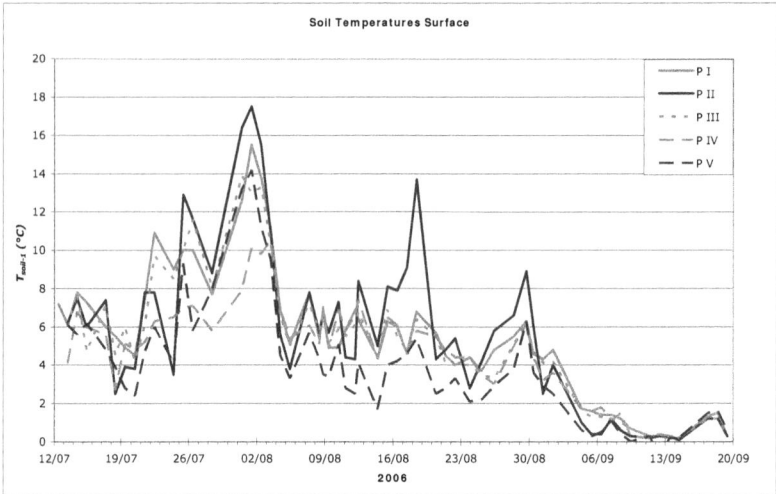

Figure 4-9: Soil surface temperatures T_{soil-1} of micro-sites P I to P V.

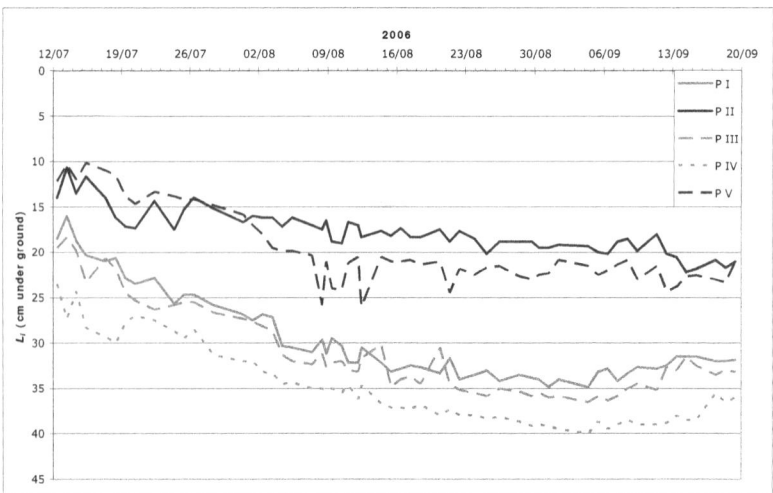

Figure 4-10: Thickness of the active layers L_i of micro-sites P I to P V.

4.2   Methane Fluxes

4.2.1   Polygon micro-sites

4.2.1.1   Micro-site P I

On average, the daily methane flux was about 100 mg m^{-2} d^{-1} for the summer months July and August, disregarding the dates with extreme environmental conditions, e.g. July 24th and August 1st, where standard deviations ranged were ± 300 mg m^{-2} d^{-1} and 150 mg m^{-2} d^{-1}, respectively. At the beginning of September fluxes decreased to less than 50 mg m^{-2} d^{-1} (Figure 4-11), closely following fluctuations in air temperature. After a short period of stormy and wet conditions during the first days of September CH$_4$ flux increased again before it fluctuated around an average of 10..20 mg m^{-2} d^{-1} from September 10th...15th. At this time, temperatures decreased to about -8 °C and snow began to addumulate. Water surfaces froze to a thickness of up to 8cm and the upper soil layers were frozen solid.

Peak fluxes on July 24th and August 1st conincided with high temperatures and falling water levels. Figure 4-11 shows the high spatial variability that is common even within the micro-site, as the standard deviation of some averaged flux values reach quite wide ranges compared to the averaged standard error of the single plot fluxes. The multiple linear regression analysis provides the following model equation:

$$
\begin{aligned}
F_{\mathrm{PI}}(t) \;=\;\; & -0.76 \cdot mg \cdot m^{-2} \cdot d^{-1} + 11.35 \cdot mg \cdot m^{-2} \cdot d^{-1} \cdot {}^{\circ}C^{-1} \cdot T_{Soil-1}(t) \\
& + 4.95 \cdot mg \cdot m^{-2} \cdot d^{-1} \cdot 100 \mu mol^{-1} \cdot m^{2} \cdot d^{-1} \cdot PAR(t)
\end{aligned}
\tag{5}
$$

with an adjusted coefficient of determination $\overline{R}^{2} = 0.70$.

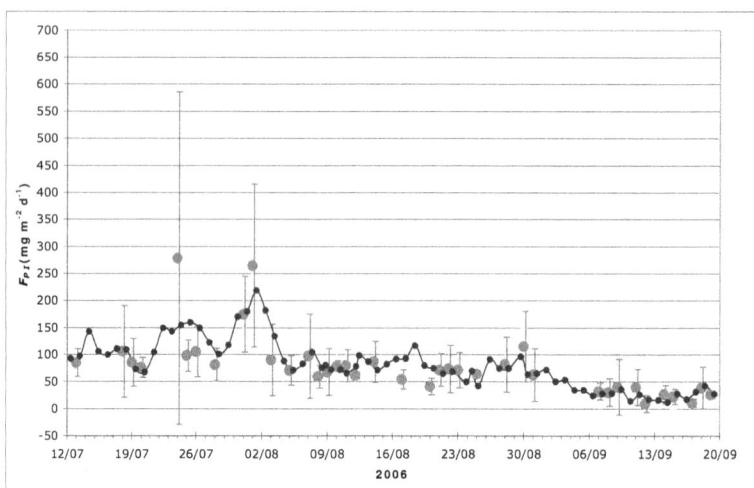

Figure 4-11: **Methane fluxes at micro-site P I** – red dots indicate measured flux values, dark blue dots and line indicate modelled flux values, blue error bar indicates standard deviation of averaged values, green error bar indicates mean standard error of measured values.

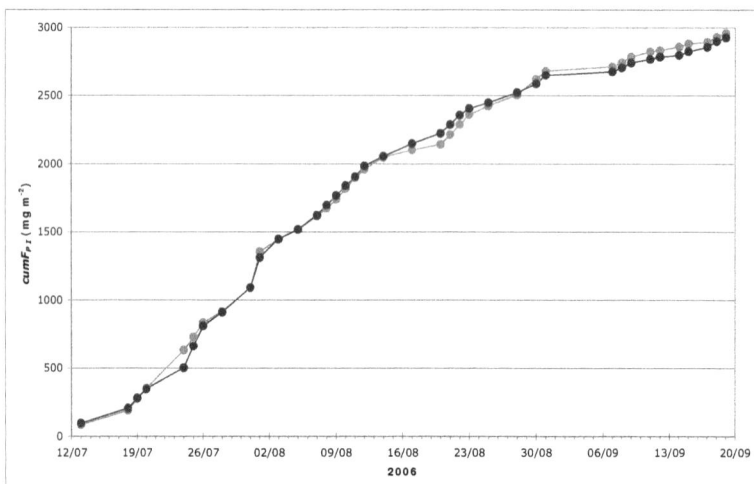

Figure 4-12: **Cumulated methane fluxes at micro-site P I** – red indicates measured fluxes, blue indicates modelled fluxes.

The cumulative measured CH_4 flux $cumF_{PI}$ was 2960 mg m^{-2}. The cumulative modelled flux of the same days amounts to 2928 mg m^{-2}. Thus, the model reflects 98.93% of the actually measured methane fluxes at micro-site P I. Figure 4-12 shows the cumulated fluxes of both measured and modelled methane fluxes at micro-site P I. Both graphs show simultaneous slopes over time. The cumulative modelled flux over the entire period from July 12th..September 19th is 5922 mg m^{-2}.

4.2.1.2   Micro-site P II

Very low methane concentrations in the closed chamber system at this site frequently caused the analyzer to reach its detection limit, resulting in very noisy data and a high exclusion rate during flux calculation (Figure 4-13). Fluxes that could be calculated were very low throughout the campaign and rarely exceeded 10 mg m^{-2} d^{-1}. That is about 10% of the fluxes from low centre polygon micro-sites. No seasonal course is evident from the data and no statistically significant correlation with any of the observed environmental parameters was found.. Therefore, a multiple linear regression model could not be set up for this micro-site. Gaps in the time series were filled with monthly averaged flux values. This approximation takes into account that small positive fluxes likely exist, even though they cannot be resolved by the analyzer. In total the cumulative measured flux is 409 mg m^{-2}. The gap-filled cumulative flux for the entire period is 730 mg m^{-2}.

4.2.1.3   Micro-site P III

In general, the seasonal course of CH_4 flux at micro-site P III was similar to the one at micro-site P I. Again, the daily CH4 flux was about 100 mg m^{-2} d^{-1} for the summer months July and August, disregarding the dates with extreme environmental conditions, e.g. July 22nd, 24th and August 1st and as a smaller extreme August 17th. Variability on these dates was large, with standard deviations of ± 150 mg m^{-2} d^{-1}, 300 mg m^{-2} d^{-1}, 250 mg m^{-2} d^{-1} and 70 mg m^{-2} d^{-1}, respectively. At the beginning of September fluxes decreased to below 50 mg m^{-2} d^{-1} (Figure 4-14). Again, CH_4 fluxes did react sensitively on the dramatic weather changes observed in this period on Samoylov Island. After the short "bad weather" period during the first days of September CH_4 flux increased very little again to about 20..25 mg m^{-2} d^{-1} before it fluctuated marginally above 0 mg m^{-2} d^{-1} during the period of subzero temperatures and snowfall from September 10th...15th.

Extreme fluxes occurred at July 22nd, 24th, August 1st and 5th., again coinciding with high temperatures and generally "good weather" conditions. Similar to micro-site P I there is a high spatial variability even within the micro-site (Figure 4-14), as demonstrated by the large standard deviation on days with above-average methane emission The multiple linear regression analysis provides the following model equation:

$$F_{PI}(t) = -1.96 \cdot mg \cdot m^{-2} \cdot d^{-1} + 8.93 \cdot mg \cdot m^{-2} \cdot d^{-1} \cdot {}^\circ C^{-1} \cdot T_{Soil-1}(t)$$
$$+ 5.57 \cdot mg \cdot m^{-2} \cdot d^{-1} \cdot {}^\circ C^{-1} \cdot T_{Ch}(t)$$

(6)

with an adjusted coefficient of determination $\overline{R}^2 = 0.62$.

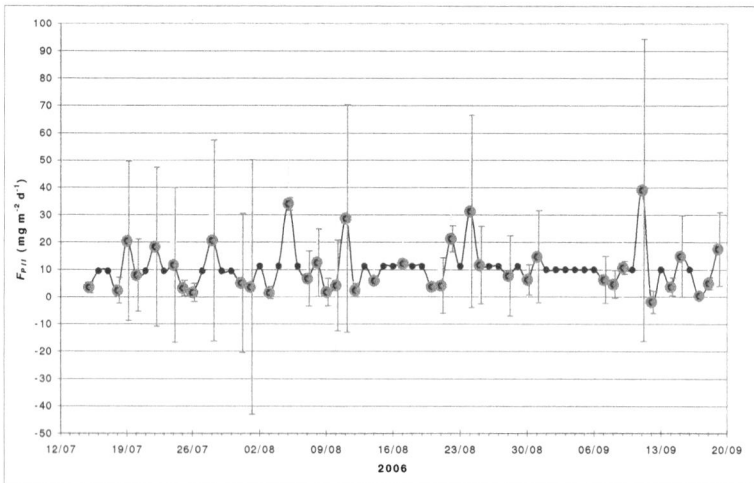

Figure 4-13: **Methane fluxes at micro-site P II** – red dots indicate measured fluxes, dark blue dots and line indicate monthly averaged fluxes, blue error bar indicates standard deviation of averaged values, green error bar indicates mean standard error of measured values.

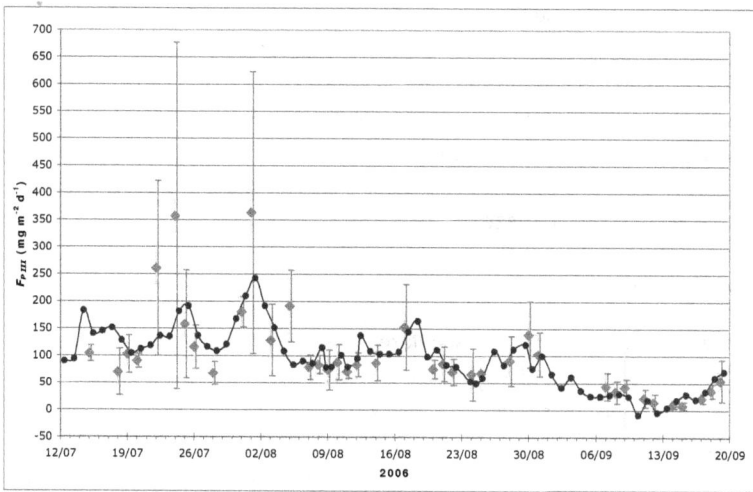

Figure 4-14: Methane fluxes at micro-site P III – red dots indicate measured flux values, dark blue dots and line indicate modelled fluxes, blue error bar indicates standard deviation of averaged values, green error bar indicates mean standard error of measured values.

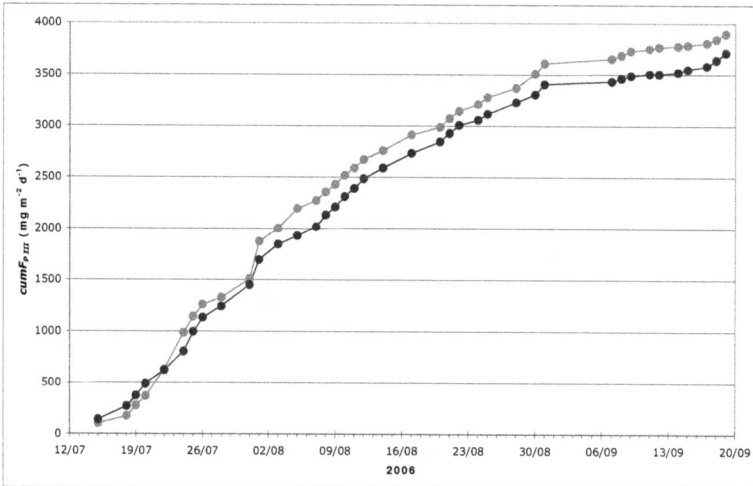

Figure 4-15: Cumulated methane fluxes at micro-site P III – indicates measured fluxes, blue indicates modelled fluxes.

- 46 -

The cumulative measured CH_4 flux $cumF_{PIII}$ was 3899 mg m^{-2} regarding the days in observation. The cumulative modelled flux of the same days was 3714 mg m^{-2}. Thus, the model reflects 95.25% of the actually measured methane fluxes at micro-site P I. Figure 4-15 shows the cumulated fluxes of both measured and modelled methane fluxes at micro-site P III. Both graphs show simultaneous slopes over time. The cumulative modelled flux over the entire period from July 12th ... September 19th was 6867 mg m^{-2}.

4.2.1.4   Micro-site P I V

The seasonal course of the methane fluxes was less pronounced at this micro-site than at micro-sites P I and P III. While average fluxes were in the same range as at the other sites, peaks and variability seem much less extreme. Standard deviations never exceeded ± 100 mg m^{-2} d^{-1}. At beginning of September fluxes decreased to below 50 mg m^{-2} d^{-1} (Figure 4-16). After the short "bad weather" period during the first days of September, CH_4 flux increased very little again to about 45..50 mg m^{-2} d^{-1} before it fluctuated marginally above 0 mg m^{-2} d^{-1} during the frost period from September 10th...15th.

The multiple linear regression analysis provides the following model equation:

$$F_{PI}(t) = 288.43 \cdot mg \cdot m^{-2} \cdot d^{-1} + 7.99 \cdot mg \cdot m^{-2} \cdot d^{-1} \cdot {}^{\circ}C^{-1} \cdot T_{Soil-1}(t)$$

$$+ 2.71 \cdot mg \cdot m^{-2} \cdot d^{-1} \cdot 100 \mu mol^{-1} \cdot m^2 \cdot d^{-1} \cdot PAR(t) \qquad (7)$$

$$- 6.27 \cdot mg \cdot m^{-2} \cdot d^{-1} \cdot 100 m^{-1} \cdot W_{col}(t)$$

with an adjusted coefficient of determination $\overline{R}^2 = 0.78$. Here, the model consists of three flux driving parameters instead of two for the models of micro-sites P I and P III, which results in the highest adjusted coefficient of determination $\overline{R}^2$.

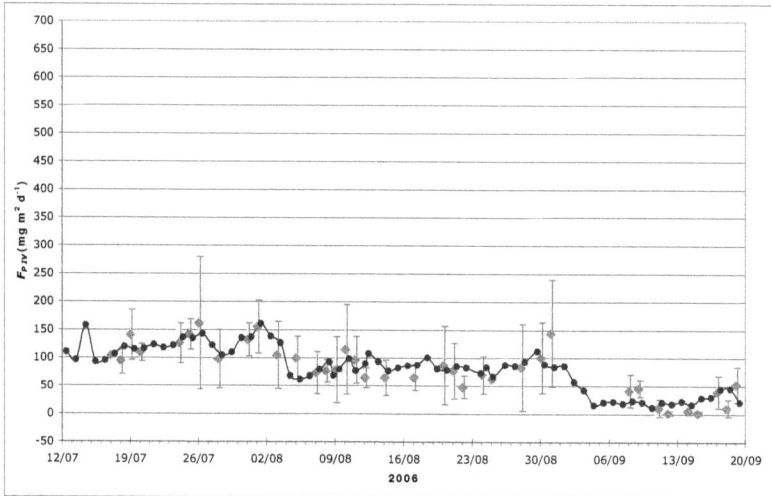

Figure 4-16: **Methane fluxes at micro-site P I V** – red dots indicate measured fluxes, dark blue dots and line indicate modelled fluxes, blue error bar indicates standard deviation of averaged values, green error bar indicates mean standard error of measured values.

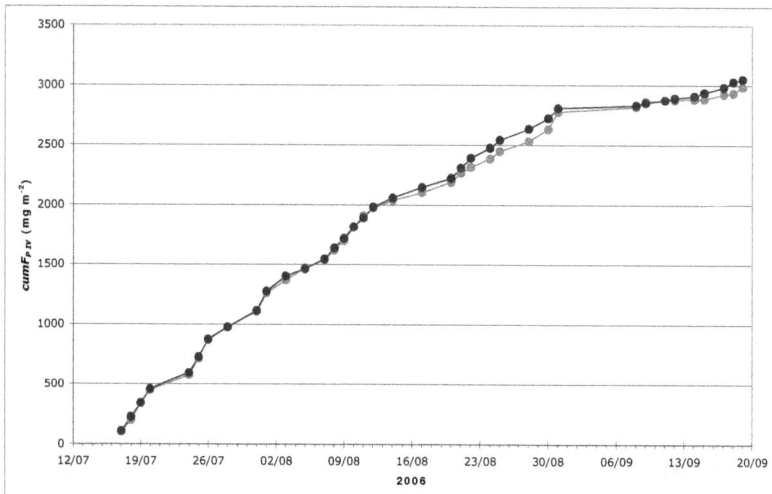

Figure 4-17: **Cumulated methane fluxes at micro-site P I V** – red indicates measured fluxes, blue indicates modelled fluxes.

The cumulative measured CH_4 flux $cumF_{PIV}$ was 2988 mg m^{-2}. The cumulative modelled flux of the same days was 3051 mg m^{-2}. Thus, the model overestimates the methane emission of this micro-site by 2.10%. Figure 4-17 shows the cumulated fluxes of both measured and modelled methane fluxes at micro-site P IV. Again, both graphs show simultaneous slopes over time. The cumulative modelled flux over the entire period from July 12th...September 19th was 5917 mg m^{-2}.

4.2.1.5   Micro-site P V

Similar to micro-site P II, only very low fluxes of about 5...10 mg m^{-2} d^{-1} were calculated for the polygon rim (Figure 4-18). Again, the detection of the Innova 1412 was reached frequently. No clear seasonal course was visible in the flux time series and no statistically significant correlation with any of the observed environmental parameters could be identified. With no applicable multiple linear regression model, gaps were filled with monthly averaged flux values. In total, the measured fluxes were 188 mg m^{-2}. The gap-filled cumulative flux for the entire period is estimated at 343 mg m^{-2}.

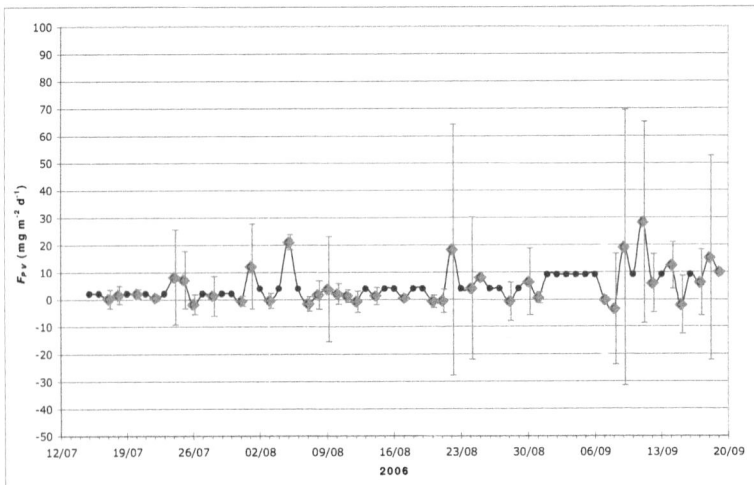

Figure 4-18: **Methane fluxes at micro-site P V** – red dots indicate measured fluxes, dark blue dots and line indicate monthly averaged fluxes, blue error bar indicates standard deviation of averaged values, green error bar indicates mean standard error of measured values.

4.2.2   Crack/trough micro-sites

Figure 4-19 shows the non-processed results of the long-term measurements conducted on the water-filled cracks and troughs which were situated between the polygon micro-sites of interest. A rough estimate of fluxes, calculated by linear regression, is given below:

- micro-site W IIb: 0.12 mg m² d^{-1},

- micro-site W IIIb: 0.22 mg m² d^{-1},

- micro-site W IIIc: 0.51 mg m² d^{-1},

- micro-site W I: 0.73 mg m² d^{-1},

- micro-site W IIa: 1.45 mg m² d^{-1}, and

- micro-site W IIIa: 2.58 mg m² d^{-1}.

Micro-sites W I, W IIa and W IIIa show significant increases of CH_4 concentration over time, indicating constant soil-atmosphere directed CH_4 flux. These three micro-sites were those, where vegetation emerges from the water surface (cmp. Chapter 3.2.2). The three remaining micro-sites W IIb, W IIIb, and W IIIc were sites of non-vegetated water surface and the concentration change over time at these sites was much less pronounced than at sites with vegetated water surface. In order to get a rough idea of the potential methane release by ebullition from these sites, CH_4 mobilization was forced by disturbing the bottom sediment in the cracks using a metal probe. The sites reacted immediately on this disturbance and CH_4 concentration in the chamber headspace increased rapidly by about one order of magnitude (Figure 4-19).

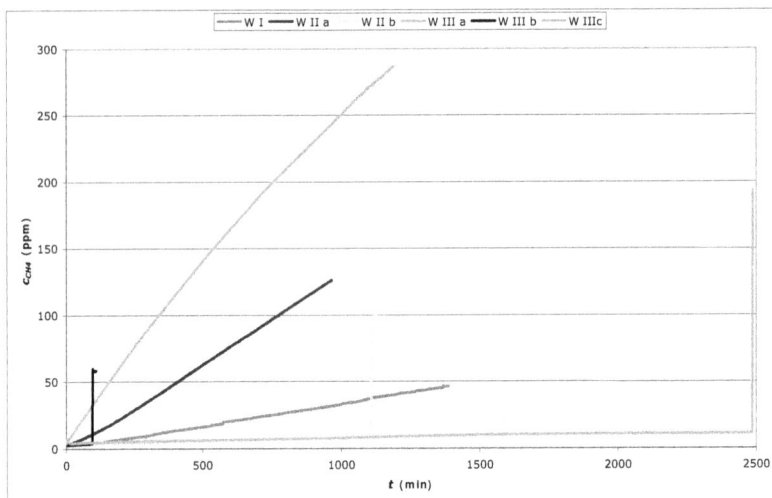

Figure 4-19: Evolution of chamber headspace c_{CH4} in long-term measurements.

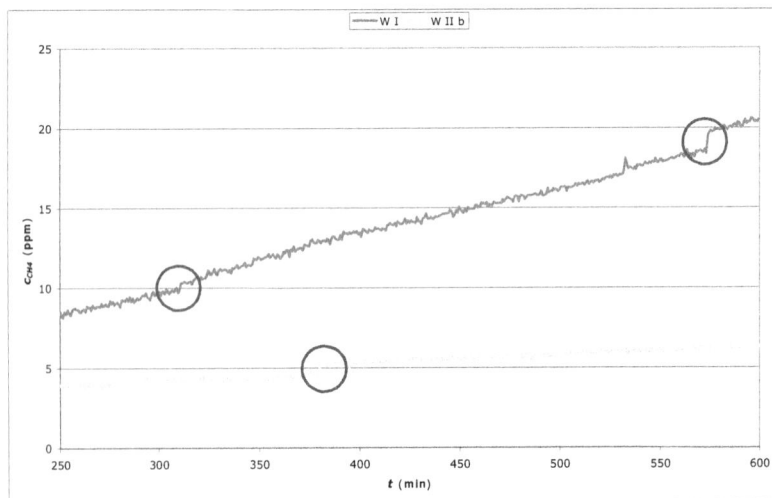

Figure 4-20: Methane ebullition events in micro-sites W I and W IIb – the figure shows a detail of Figure 4-19.

Natural ebullition was detected at micro-sites W I and W IIb. Zooming into the concentration time series, three distinct events could be identified, where a sudden increase in concentration was preceded and followed by a steady increase. (Figure 4-20). The measurement setup excludes the possibility to determine if one event consisted of one big or several small gas bubbles.

Regarding the first ebullition event at micro-site W I, 2.39 mg m^{-2} CH$_4$ were released within 90 seconds. At micro-site W IIb the CH$_4$ release amounts to 4.40 mg m^{-2}. The second ebullition event at micro-site W I is clearly identified as the largest one of the three: 8.65 mg m^{-2} CH$_4$ altered the concentration within 45 seconds under the deployed floating chamber. The ebulltion events were excluded regarding the estimated fluxes of micro-sites W I and W IIb. That is as the ebullition as extreme events altered the averaged fluxes considering the diurnal perspective.

5   Discussion

5.1   Seasonal Dynamics

The average methane flux (averaged over all micro-sites) was 57 mg m^{-2} d^{-1}. The average flux recorded by the eddy covariance system during the same period was much lower at 20 mg m^{-2} d^{-1} [Sachs et al., 2007].

Seasonal dynamics of plot-based methane fluxes and eddy covariance methane fluxes also differed clearly: fluxes from polygon centres showed a clear seasonal course with highest fluxes in July and gradually decreasing emission towards the end of the season. This seasonal course was much less visible in the eddy covariance data [Sachs et al., 2007]. Distinct emission peaks at P I and P III occured on July 22nd ... July 24th and August 1st with methane emission exceeding 250 mg m^{-2}. These maxima coincided with maxima of PAR and air temperature T_{air} (cmp. Figure 4-3 and Figure 4-4) and the warmest soil surface temperature T_{Soil-1} was measured on August 1st. Thus, at polygon centre micro-sites, maximum methane emissions generally appeared to be associated with "good weather", which usually also results in falling water levels. Peaks in the eddy covariance time series, however, were associated with increased wind speed, decreased atmospheric pressure, and strong precipitation events. While fluxes from rim and high centre micro-sites were generally very low, peaks occured here during 'bad weather" conditions at the beginning of August, a few days after the peaks observed at the low centre sites. The peaks at the polygon rim and high centre sites coincided with a beginning of August peak in methane emission recorded by the eddy covariance system.

These differences in the seasonal dynamics can be explained best by the very different hydrological conditions of the investigated micro-sites: in the wet polygon centres, water levels were always above the soil surface or at least the criterion of water saturation was fulfilled. Here, higher water levels could lead to decreased methane emission, as more of the vegetation becomes submerged and plant-mediated transport decreases. Hence, warm weather and falling water levels could increase emissions. At "drier" micro-sites, on the other hand, storm systems with strong precipitation events lead to a temporary increase in anaerobic soil volume and an increase in methane production, while lower temperatures have a negative effect on the activity of methane oxidizing microbes.

In consequence, these observations lead to the hypothesis that polygon rim and high centre micro-sites must be dominant in the eddy covariance footprint.

It should be noted, that during a long-term measurement of a water-filled trough in the night of August 31st...September 1st the floating chamber got disconnected from the sensor unit in a snowstorm, causing the analyzer to suck water into the sample cell. It took several days to dry the analyzer, and when measurements were resumed, weather conditions had changed dramatically. Temperatures dropped below 0 °C and snow started to accumulate. Thus, it cannot completely be ruled out that the lower fluxes measured in September were caused by damage to the analyzer. However, the measured fluxes decreased with freezing of the top soil and lower temperatures, and increased as expected at the end of the measurement period, when soil thawed due to warmer temperature. It is likely, that snow cover and low temperatures did indeed cause the low 50...70% reduction in methane fluxes observed in September (cmp. Figure 4-11,Figure 4-14 and Figure 4-16).

5.2 Controls on methane fluxes

Due to the very low fluxes from P II and P V, statistically verified controls on methane flux could only be identified for the water saturated micro-sites P I, P III, and P IV. All three multiple regression models include soil surface temperature T_{Soil-1} as a significant parameter (cmp. eqs. (5), (6) and (7)). Since the temperature dependence of methane fluxes is usually due to the temperature dependence of microbial processes, a lower soil temperature would have been expected to be a significant control on methane fluxes. However, cold permafrost curbs temperature fluctuations in deeper soil layers and relatively stable temperatures are mathematically less suitable for explaining variable fluxes than air or soil surface temperatures, which show more fluctuation. Comparing the soil surface temperature T_{soil-1} with the seasonal temperature course of lower soil layers, the temperatures from underneath can be seen as noisy measure of the amplified T_{soil-1} (cmp. Figure 4-9 with Figure 7-1 and Figure 7-2). Thus, the soil surface temperature T_{soil-1} should not be interpreted as a steering parameter but could be seen as a significant indicator for the activity of the methanogenic bacteria in water saturated polygons.

At P III, chamber temperature T_{Ch} was the second significant parameter in the multiple linear regression model (cmp. eq. (6) in Chapter 4.2.1.3), although it can be expected to be directly correlated with soil surface temperature, air temperature T_{air}, and PAR. Nevertheless, as chamber temperature was more of an artefact than a true environmental parameter, this underscores the fact that the chamber method itself influenced the measurand. i.e. the methane flux.

At micro-site P IV three significant parameters were identified (cmp. eq. (7) in Chapter 4.2.1.4). In addition to soil surface temperature T_{Soil-1} and PAR the model includes the calculated parameter water column W_{col}. As the water saturation and thus an anaerobic layer are required for the methane production this does not appear as remarkable. Surprisingly though, this parameter has a negative coefficient, i.e. a higher water column results in a lower methane flux. This becomes understandable when considering that the water level at P IV was the highest throughout the season. As soon as standing water above the surface is present, this hampers methane flux as less and less plant-mediated transport can occur, when water levels are too high.

PAR was another significant control on methane emissions at micro-sites P I and P III (cmp. eqs. (5) and (6)). As PAR is known to control plant photosynthesis, it might be interpreted as an indicator for plant activity, here the aerenchym activity, which enables fast CH_4 efflux from the plant roots to the atmosphere [Kutzbach, 2000]. Certainly, PAR causes warming of the irradiated surfaces and is therefore directly correlated or a potential substitute for surface temperature. Analogue to the effect of direct solar radiation on chamber temperature, it can be assumed that PAR increases the temperature of the upper water layer underneath a chamber. Henry's law states that at a constant temperature, the amount of a given gas dissolved in a given type and volume of liquid is directly proportional to the partial pressure of that gas in equilibrium with that liquid. The proportionality is expressed by the Henry constant:

$$p_{gas} = k_{H,pc} \cdot c_{aq} \tag{8}$$

where p_{gas} is the partial pressure of the gas, c_{aq} the concentration in the solution, and $k_{H,pc}$ is the Henry constant. With increasing water temperature underneath a chamber the equilibrium concentration decreases and dissolved methane is mobilized. If this speculation is true, the localized effect of PAR on water temperatures and methane emission would be another chamber-induced artefact. The solubility of CH_4 in H_2O is low compared to other gases. But in fact, the solubility of 35 ml l^{-1} (equivalent to 35,000 ppmv) at 20 °C holds some potential to influence the atmospheric CH_4 concentration. *Spott [2003]* describes a dependency of the diffusion transfer velocity k_i from the temperature T_{water} in the water of the observed thermokarst lakes on Samoylov Island. There, transfer velocity k_i doubles within the interval of water temperature from 6 °C...12 °C. According to Fick's first law (cmp. eq. (2) in Chapter 2), this results in the assumption that at constant amount of dissolved CH_4 diffusion increases with rising water temperature.

No significant controls on methane fluxes could be identified for P II and P V. Nevertheless, the low methane emission from these sites can still be explained by the underlying microbial processes. Assuming that there is similar methane production in the anaerobic layers above the permafrost table, the aerobic layer (ventilated because of non-water saturated conditions; cmp. Figure 4-7) compensates methane production by similarly large methane consumption. However, methane production is less on drier micro-sites to begin with. This is mainly due to thinner anaerobic layer (=water-saturated layer) This thin anaerobic layer is close to the permafrost table, where temperatures are low and cause methanogenic activity to be lower than in warmer soil layers further away from the permafrost table, which are still water-saturated and (i.e. anaerob) in polygon centres. While most of the methane is oxidized in the thick overlying aerobic layer, small amounts can be released either by bypassing methane oxidizers or when the delicate balance between production and consumption occasionally shifts towards production.

5.3   Crack/trough micro-sites

During the long-term measurements on the polygon cracks and troughs, the three main methane emission paths ebullition, vegetation-mediated transport, and diffusion could be identified at least partially.

Three ebullition events were recorded directly and are clearly visible in the continuous concentration data from the crack and trough micro-sites (cmp. Figure 4-20). Gas bubbles raised the CH_4 concentration under the deployed floating chamber by releasing 2.39 mg m^{-2}... 8.65 mg m^{-2} into the chamber headspace. However, generalization of these observations is difficult due to the temporally and spatially random nature of ebullition events. In general, if the sum of partial pressures reaches a higher value than the one of hydrostatic pressure, bubbles grow and then emit to the atmosphere. Thus, the process of bubble-development is not trivial in reconstruction. In addition, this process can be hindered and interrupted, respectively, as plant-mediated transport runs continuously and thus removes most of the gas before its accumulation [*Chanton & Whiting, 1995* cited in *Spott, 2003*].

Furthermore, *Spott* [*2003*] showed that plant–mediated efflux by emerging vegetation reached 79..98% of measured CH_4 fluxes on water surfaces. This is in agreement with the present study, where micro-sites W IIIa, W IIa, and W I with emerging vegetation showed higher fluxes compared to the ones with non-vegetated water surface. Since no other flux path could clearly be identified at the micro-sites with non-vegetated water surface, it can be assumed that fluxes were based on diffusion. In addition, one can assume that the measured fluxes at the micro-sites with emerging vegetation are the sum of both plant-mediated efflux and diffusion. Taking this into consideration, the pure vegetation-mediated efflux can be estimated by the difference between the CH_4 flux from a micro-site with emerging vegetation and one without emerging vegetation from the same crack or trough. Thus, the proportion of vegetation-mediated transport from the measured total CH_4 efflux was approximately 92%, 80% and 92%, respectively, at micro-site W IIIb, W IIIc, and W IIb.

It should be kept in mind that the presented results for the ad hoc long-term measurements on crack and trough micro-sites are one-time measurements only. Thus, they have to be regarded as preliminary examinations and not as statistically validated conclusions. Nevertheless, the presented results support earlier findings by *Spott* [*2003*].

5.4 Conclusions

- The data set collected during this study is unique with regard to it's extend, the investigation area, and it's further potential for integrating flux measurements on different spatial scales. It represents a detailed record of high-quality methane emission data from several soil-vegetation units, which have previously not been available for this area.

- The investigation area in the centre of Samoylov Island is a source of methane, but not as strong as source as might be expected from a "wet" tundra. This is due to the high spatial variability of the methane emissions.

- Methane fluxes differ by an order of magnitude between polygons with low-centre character and polygons with high-centre character or polygon rims, respectively. However, accurate flux measurements at micro-sites with extremely low fluxes require analyzers with a lower detection limit and better precision, in order to increase the signal to noise ratio.

- Simple multiple linear regression models described the measured fluxes from wet polygon centres reasonably well. Using this approach, soil surface temperature T_{Soil-1} was identified as a main environmental control on methane emissions from water-saturated micro-sites, along with other (partly correlated) parameters that varied between micro-sites. Although directly measured water levels were not found to be significant parameters, the height of the water column, which can change due to increasing active layer depth, even when the water level measured from the surface remains stable, was a significant control at micro-site P IV. This parameter also has important implications with regard to whether water is present above the soil surface, and should be investigated in more detail in future studies due to its potential duplicate function as both methane flux buffer and methane reservoir. The significance of chamber temperature as a control on methane emission from micro-site P III, which was the strongest source of methane, raises some questions regarding the reliability of closed chamber measurements, even when closure times are kept short, if climatic conditions during chamber closure can not be controlled and kept near ambient conditions.

- At the crack/trough micro-sites, ebullition of methane was detected in "real-time" due to the continuous sampling. The importance of plant-mediated transport of methane from thermokarst features and ponds was confirmed and the potential for significant methane release by ebullition from bottom sediments was demonstrated. Future investigations of methane fluxes on Samoylov Island should pay closer attention to water bodies in the eddy covariance footprint as well as the temporal and spatial distribution of methane release by ebullition.

6 References

ACIA, 2004: *Impacts of a Warming Arctic*: Arctic Climate Impact Assessment, Cambridge University Press, New York, USA.

ALABYAN, A. M., CHALOV, R. S., KOROTAEV, V. N., SIDORCHUK, A. Y., ZAITSEV, A. A., 1995: *Natural and technigenic water and sediment supply to the Laptev Sea*, Reports on Polar Research 176: 265-271.

ALM, J., SHURPALI, N. J., TUITTILA, E.-S., LAURILA, T., MALJANEN, M., SAARNIO, S., AND MINKKINEN, K. , 2007: *Methods for determining emission factors for the use of peat and peatlands – flux measurements and modelling*, Boreal Environmental Research, 12, 85–100.

ALM, J., TALANOV, A., SAARNIO, S., SILVOLA, J., IKKONEN, E., AALTONEN, H., NYKÄNEN, H., AND MARTIKAINEN, P. J. , 1997: *reconstruction of the carbon balance for microsites in a boreal oligotrophic pine fen*, Finland, Oecologia, 110, 423–431.

ARE, F. E., REIMNITZ, E., 2000: *An overview of the Lena River Delta setting: geology, tectonics, geomorphology and hydrology*, Journal of Coastal Research 16(4): 1083-1093.

BILLINGS, 1987, W. D., 1987: *Carbon balance of Alaskan tundra and taiga ecosystems: past, present and future*, Quaternary Science Review 6: 165-177.

BOIKE, J., HINZMAN, L., OVERDUIN, P. P., ROMANOVSKY, V., IPPISCH, O., ROTH, K., 2003: *A comparison of snow melt at three circumpolar sites: Spitsbergen, Siberia, Alaska*, Poster at the 8th International Conference on Permafrost, July 21st...25th, Zürich, Switzerland.

BOIKE, J., 2006: personal correspondence, not published data.

BUBIER, J., CRILL, P., AND MOSEDALE, A., 2002: *Net ecosystem CO_2 exchange measured by autochambers during the snow-covered season at a temperate peatland*, Hydrological Processes, 16, 3667–3682.

BURROWS, E. H., BUBIER, J. L., MOSEDALE, A., COBB, G. W., AND CRILL, P. M., 2004: *Net Ecosystem exchange of carbon dioxide in a temperate poor fen: a comparison of automated and manual chamber techniques*, Biogeochemistry, 76, 21–45.

CHANTON, J. P. AND WHITING, G. J., 1995: *Trace gas exchange across the air-water interface in freshwater and coastal marine environments: ebullition and transport by plants*. In: MATSON, P. A. & R. C. HARRISS (Eds.), *Biogenic trace gases: Measuring emissions from soil and water*, Blackwell Science. Oxford.

CHRISTENSEN, T., 1993: *Methane emission from arctic tundra*, Biogeochemistry 21: 117-139.

CHRISTENSEN, T. R., JONASSON, S., MICHELSEN, A., CALLAGHAN, T. V., AND HAVSTRÖM, M., 1998: *Environmental controls on soil respiration in the Eurasian and Greenlandic Arctic*, Journal of Geophysical Research, 5 103(D22), 29 015–29 021.

CONRAD, R., 1989: *Control of methane production in terrestrial ecosystems*. In: ANDREAE, M.O., SCHIMEL, D.S. (Eds.). *Exchange of Trace Gases between Terrestrial Ecosystems and the Atmosphere*. Chichester: John Wiley and Sons. pp. 39-58.

EMBELTON, C. AND KING, C.A.M., 1975: Periglacial geomorphology. London.

ETHERIDGE, D. M., STEELE L.P., FRANCEY R.J., LANGENFELDS, R. L.: *Atmospheric methane between 1000 A.D. and present: Evidence of anthropogenic emissions and climatic variability*, Journal of Geophysical Research 103: 15979-15993.

EUGSTER, W., ROUSE, W. R., PIELKE, R. A. SENIOR; MCFADDEN, J. P., BALDOCCHI, D. D. KITTEL, T. G., CHAPIN, F. S. III, LISTON, G. E., VIDALE, P. L., VAGANOV, E., CHAMBERS, S., 2000: *Land- atmosphere energy exchange in Arctic tundra and boreal forest: available data and feedbacks to climate*, Global Change Biology 6(S1): 84-115.

FRAHM, J.-P., WOLFGANG FREY, W., DÖRING, J., 2004: *Moosflora*. 4., neu bearbeitete und erweiterte Auflage (UTB für Wissenschaft, Band 1250). Ulmer, Stuttgart

FRENCH, H. M. 1996: *The Periglacial Environment*. Longman. Harlow.

FRITSCHE, W. 2002: *Mikrobiologie*, Spektrum Akademischer Verlag. Berlin.

GATES, D. M., 1980: *Biophysical Ecology*, Springer-Verlag, New York, 611 p.

GORHAM, E., 1991: *Northern peatlands: role in the carbon cycle and probable responses to climatic warming*. Ecological Applications 1: 182-195.

GOULDEN, M. L. AND CRILL, P. M. , 1997: *Automated measurements of CO_2 exchange at the moss surface of a black spruce forest*, Tree Physiology, 17, 537–542.

GRIGORIEV, M. N., 1993: *Cryomorphogenesis of the Lena River mouth area*, SB RAS, Yakutsk, Russia, 176 pp (in Russian).

HAPPELL, J. D., CHANTON, J.P., WHITING, G.J., SHOWERS, W., 1993: *Stable Isotopes as tracers of methane dynamics in Everglades marshes with and without active populations of methane oxidizing bacteria*, Journal of Geophysical Research 98: 14771-14782.

HASSELMANN, K., LATIF, M., HOOSS, G., AZAR, C., EDENHOFER, O., JAEGER, C. C., JOHANNESSEN, O. M., KEMFERT, C., WELP, M., WOKAUN, A., 2003: *The challenge of long-term climate change*, Science 302: 1923-1925.

HEIJMANS, M. P. D., ARP, W. J., AND CHAPIN III, F. S., 2004: *Carbon dioxide and water vapour exchange from understorey species in boreal forest*, Agricultur Forest Meteorology, 123, 135–147.

HEYER, J., 1990: *Der Kreislauf des Methans*, Berlin, Akademie-Verlag, 250 p.

HITCHCOCK, C. L., CRONQUIST, A., OWNBEY, M., THOMPSON, J. W., 1969: *Vascular Plants of the Pacific Northwest*, Part 1, Seattle, University of Washington Press. 914 p.

HMCR HYDROMETEOROLOGICAL CENTRE OF RUSSIA, 2007: Russia's Weather (URL: http://meteo.infospace.ru, March 2007).

HOLZAPFEL-PSCHORN, A., CONRAD, R., SEILER, W., 1986: *Production, oxidation and emission of methane in rice paddies*. FEMS Microbiology Ecology 31: 343-351.

INNOVA Airtech Instruments, 2006: *Instruction Manual*, LumaSense Technologies A/S - INNOVA Airtech Instruments, Ballerup, Denmark

IPCC (Intergovernmental Panel of Climate Change), 1995: *Climate Change 1995, The Science of Climate Change, Contribution of Working Group I to the Second Assessment Report of the Intergovernmental Panel on Climate Change*, Houghton J.T., Meira Filho, L.G., Callander, B.A., et al. (Eds.), Cambridge, University Press. 570 p.

IPCC (Intergovernmental Panel of Climate Change), 2001: *Climate Change 2001: Working Group I: The Scientific Basis: C.1 Obsereved Changes in Globally Well-Mixed Greenhouse Gas Concentrations and Radiative Forcing* (URL: http://www.grida.no/climate/ipcc_tar/wg1/016.htm, January 2007)

IVANOV, V. V. AND PISKUN, A. A., 1995: *Distribution of river water and suspended sediments loads in the river deltas of rivers in the basins of the Laptev and East-Siberian Seas*. In: KASSENS et al. (Eds.) *Land-Ocean systems in the Siberian Arctic: Dynamics and history*, Lecture notes in earth science, Springer, Berlin, pp 239-250

JENSEN, L. S., MUELLER, T., TATE, K. R., ROSS, D. J., MAGID, J., AND NIELSEN, N. E. , 1996: *Soil surface CO_2 flux as an index of soil respiration in situ: a comparison of two chamber methods*, Soil Biology Biochemistry, 28, 1297–1306

KING, G. M. 1990. *Dynamics and controls of methane oxidation in a Danish wetland sediment*. FEMS Microbiol Ecol 74: 309-324.

KNOBLAUCH, C., 1994: *Bodenkundlich-mikrobiologische Bestandsaufnahme zur Methanoxidation in einer Flussmarsch der Tide-Elbe*, Diploma thesis, Institut für Bodenkunde, Universität Hamburg.

KOTLYAKOV, V. AND KHROMOVA, T., 2002: *Permafrost, Snow and Ice*. In: STOLBOVOI, V. AND MCCALLUM I (Eds.) CD-ROM *Land Resources of Russia*, International Institute of Applied Systems Analysis and the Russian Academy of Science, Laxenburg, Austria.

KUTZBACH, L., 2000: *Die Bedeutung der Vegetation und bodeneigener Parameter für die Methanflüsse in Permafrostböden*, Diploma thesis, Institut für Bodenkunde, Universität Hamburg

KUTZBACH, L., 2006: *The Exchange of Energy, Water and Carbon Dioxide between Wet Arctic Tundra and the Atmosphere at the Lena River Delta*, Reports on Polar and Marine Research, 541, 157 pp., Alfred Wegener Institute, Bremerhaven, Germany.

KUTZBACH, L., SACHS, T., GIEBELS, M., NYKÄNEN, H., SHURPALI, N. J., MARTIKAINEN, P. J., ALM, J., WILMKING, M., 2007: *CO_2 flux determination by closed-chamber methods can be seriously biased by inappropriate application of linear regression*, Biogeosciences Discussions 4: 2279-2328.

LAINE, A., SOTTOCORNOLA, M., KIELY, G., BYRNE, K. A., WILSON, D., AND TUITTILA, E.-S., 2006: *Estimating net ecosystem exchange in a patterned ecosystem: Example from blanket bog*, Agricultur Forest Meteorology, 18, 231–243.

LOVELAND, T. R., REED, B. C., BROWN, J. F., OHLEN, D. O., ZHU, Z., YANG, L., MERCHANT, J. W., 2000: *Development of a global land cover characteristic database and IGBP DISC over from 1 km AVHRR data*, Journal of Remote Sensing, 21(6): 1303-1330.

MALJANEN M., MARTIKAINEN P. J., WALDEN J., SILVOLA J., 2001: *CO_2 exchange in an organic field growing barley or grass in eastern Finland*, Global. Change Biology, 7, 679–692.

MASSACHUSETTS DIVISION OF FISHERIES & WILDLIFE, 2004: *Natural Heritage Endangered Species Program* (URL: http://www.state.ma.us/dfwele/dfw/nhesp)

MATSON, P. AND HARRISS, R. (Eds.) , 1995: *Biogenic Trace Gases: Measuring Emissions from Soil and Water*, Blackwell Science, Oxford

MATTHEWS, E., 1983: *Global vegetation and land use: New high-resolution data base for climate studies*, Journal of Climate and Applied Meteorology 22: 474-487.

MOBERG, A., SONECHKIN, D. M., KOLMGREN, K., DATSENKO, N.M., KARLÉN, W., 2005: *Highly variable Northern hemisphere temperatures reconstructed from low- and high-resolution data*. Nature 433:613-617.

OECHEL, W. C., HASTINGS, S. J., VOURLITIS, G. L., JENKINS, M., RIECHERS, G., GRULKE, N., 1993: *Recent change of Arctic ecosystems from a net carbon dioxide sink to a source*. Nature 361: 520-523.

OECHEL, W. C., VOURLITIS, G. L., BROOKS, S., CRAWFORD, T. L., DUMAS, E., 1998: *Intercomparison among chamber, tower, and aircraft net CO_2 and energy fluxes measured during the Arctic System Science Land-Atmosphere-Ice Interactions (ARCSS-LAII) Flux Study*, Journal of Geophysical Research, 103(D22), 28 993–29 003

OECHEL, W. C., VOURLITIS, G. L., HASTINGS, S. J., ZULUETA, R. C., HINZMAN, L., AND KANE, D. , 2000: *Acclimation of ecosystem CO_2 exchange in the Alaskan Arctic in response to decadal climate warming*, Nature, 406, 978–981

PARSONS M, ZHANG T. 2003: CAPS: *The Circumpolar Active-Layer and Permafrost System*, Version 2.0. National Snow and Ice Data Center/World Data Center for Glaciology, Boulder, CO. 3 Compact Disks

PETERSON, B. J., HOLMES, R. M., MCCLELLAND, J. W., VÖRÖSMARTY, C. J., LAMMERS, R. B., SHIKLOMANOV, A. I., SHIKLOMANOV, I. A., RAHMSTORF, S., 2002: *Increasing river discharge to the Arctic Ocean*, Science 298: 2171-2173.

PING, C. L., BOCKHEIM, J. G., KIMBLE, L. M., MICHAELSON, G. J., WALKER, D. A., 1998: *Characteristics of cryogenic soils along a latitudinal transect in arctic Alaska*, Journal of Geophysical Research 103(D22): 28917-28928.

POLYAKOV, I. V., BEKRYAEV, R. V., ALEKSEEV, G. V., BHATT, U. S., COLONY, R. L., JOHNSON, M. A., MASKSHITAS A. P., WALSH, D., 2003: *Variability and trends of air temperature and pressure in the maritime Arctic 1875-2000*, Journal of Climate 16: 2067-2077.

POST, W. M., EMANUEL, W. R., ZINKE, P. J. STANGENBERGER, A. G., 1982: *Soil carbon pools and world life zones*, Nature 298: 156-159.

PUMPANEN, J., ILVESNIEMI, H., PERÄMÄKI, M., AND HARI, P., 2003: *Seasonal patterns of soil CO_2 efflux and soil air CO_2 concentration in a Scots pine forest: comparison of two chamber techniques*, Global Change Biology., 7, 371–382.

RACHOLD, V., GRIGORIEV, M. N., ARE, F. E., SOLOMON, S., REIMNITZ, E., KASSENS, H., ANTONOW, M., 2000: *Coastal erosion vs. riverine sediment discharge in the Arctic shelf sea*, International Journal of Earth Science 89: 450-460.

RACHOLD, V., GRIGORIEV, M. N., BAUCH, H. A., 2002: *An estimation of the sediment budget in the Laptev Sea during the last 5,000 years*, Polarforschung 70: 151-157.

RAHMSTORF, S., ARCHER, D., EBEL, D. S., EUGSTER, O., JOUZEL, J., MARAUN, D., NEU, U., SCHMIDT, G.A. SEVERINGHAUS, J., WEAVER, A. J., ZACHOS, J., 2004: *Cosmic rays carbon dioxide and climate*, EOS 85(4): 38-41.

RETH, S., GÖDECKE, M., AND FALGE, E. , 2005: *CO_2 efflux from agricultural soils in eastern Germany – comparison of a closed chamber system with eddy covariance measurements*, Theoretical and Applied Climatology, 80, 105–120.

ROSHYDROMET RUSSIAN FEDERAL SERVICE FOR HYDROMETEOROLOGY AND ENVIRONMENT MONITORING, 2007, Weather Information for Tiksi (URL: http://www.worldweather.org/107/c01040.htm, March 2007), World Meteorological Organization

ROULET, N. T., MOORE, T. R., BUBIER, J. L., LAFLEUR, P., 1992: *Northern fens: methane flux and climatic change*, Tellus 44B: 100-105

SACHS, T., WILLE, C., BOIKE, J., KUTZBACH, L., 2007: *Environmental controls on ecosystem-scale CH_4 emission from polygonal tundra in the Lena River Delta, Siberia*, Journal of Geophysical Research (under review)

SCHEFFER, F. AND SCHACHTSCHABEL, P., 1998: *Lehrbuch der Bodenkunde*, 14. Auflage, Spektrum Akademischer Verlag, Stuttgart

SCHNELLE, M., 2007: *Die spätquartäre Landschaftsentwicklung im Umfeld der Insel Arga Muora Sise im Lena-Delta, Nordost-Sibirien*, Diploma thesis, Fakultät für Physik und Geowissenschaften, Universität Leipzig.

SCHOFIELD, W. B. 1992. *Some Mosses of British Columbia*. Royal British Columbia Museum Handbook. Victoria. British Columbia. 394 pp.

SCHULTZ, J., 1995: *Die Ökozonen der Erde*. Stuttgart.

SCHWAMBORN, G., RACHOLD, V., GRIGORIEV, M. N., 2002: *Late quaternary sedimentation history of the Lena Delta*, Quaternary International 89: 119-134.

SERREZE, M. C., BROMWICH, D. H., CHAPIN, E. C., OSTERKAMP, T., DYUGEROV, M., ROMANOVSKY, V., OECHEL, W. C., MORISON, J., ZHANG, T., BARRY, R. G., 2000: *Observation evidence of recent change in the northern high-latitude environment*. Climate Change 46: 159-207.

SPOTT, O., 2003: *Frostmusterbedingte Seen der Polygonalen Tundra und ihre Funktion als Quellen atmosphärischen Methans*, Diploma Thesis, Fakultät für Physik und Geowissenschaften Universität Leipzig.

STOCKNER, T. F., SCHMITTNER, A., 1997: *Influence of CO_2 emission rates on the stability of the thermohaline circulation*, Nature 388: 862-865.

TENHUNEN, D., 1996: *Diurnal and seasonal patterns of ecosystem CO_2 efflux from upland tundra in the foothills of the Brooks Range, Alaska, USA*, Arctic and Alpine Research 28: 328-338.

TUITTILA, E.-S., KOMULAINEN, V. M., VASANDER, H., AND LAINE, J., 1999: *Restored cut-away peatland as a sink for atmospheric CO_2*, Oecologia, 120, 563–574.

VAN EVERDINGEN, R., 1998: *Multi-language glossary of permafrost and related ground-ice terms*. National Snow and Ice Data Center/World Data Center for Glaciology, Boulder, Colorado, USA.

VON STORCH, H., ZORITA, E., JONES, J.M., DIMITRIEV, Y., GONZÁLEZ-ROUCO, F., TETT, S. F. B., 2004: *Reconstructing past climate from noisy data*. Science 22(306): 679-682.

VOURLITES, G. L., OECHEL, W. C., HASTINGS, S. J., AND JENKINS, M. A., 1993: *A system for measuring in situ CO_2 and CH_4 flux in unmanaged ecosystems: an arctic example*, Functional Ecology, 7, 369–379

WALKER, H. J.,1998: *Arctic deltas*, Journal of Coastal research 14(3): 718-738.

WALTER, K. M., S. A. ZIMOV, J. P. CHANTON, D. VERBYLA, AND F. S. CHAPIN III, 2006: *Methane bubbling from thaw lakes as a positive feedback to climate warming*, Nature, 443, doi:10.1038/nature05040.

WANG, C., YANG, J., AND ZHANG, Q., 2006: *Soil respiration in six temperate forests in China, Global Change Biology*, 12(11), 2103–2114.

WANG, Z.P., DELAUNE, R. D., LINDAU, C. W., PATRICK, W. H. JUNIOR, 1993: *Methane production from anaerobic soil amended with rice straw and nitrogen fertilizers*, Fertilizer Research 33: 115-121

WEIN, N., 1999: *Sibirien*. Gotha.

WHALEN, S. C., REEBURGH, W. S., REIMERS, C. E., 1996. *Control of tundra methane emission by microbial oxidation*. In: REYNOLDS. J.F., TENHUNEN, J.D. (Eds.): *Landscape Function and Disturbance in Arctic Tundra*. Ecological Studies 120. Springer Verlag, Berlin. p. 257-274

WOOD, R. A., KEEN, A. B., MITCHELL, J. F. B., GREGORY, J. M., 1999: *Changing spatial structure of the thermohaline circulation in response to atmospheric CO_2 forcing in a climate model*, Nature 399: 572-575.

XU, M. AND QI, Y., 2001: *Soil-surface CO_2 efflux and its spatial and temporal variations in a young ponderosa pine plantation in northern California*, Global Change Biology, 7, 667–677

YANG, D. KANE, D. L., HINZMAN, L. D., ZHANG, X., ZHANG, T.,HENGCHUN, Y., 2002: *Siberian Lena River hydrologic regime and recent change, Journal of Geophysical* Research D23: 4694.

ZHANG, T., BARRY, R. G., KNOWLES, K., HEGINBOTTOM, J. A., BROWN, J., 1999: *Statistics and characteristics of permafrost and ground-ice distribution in the Northern hemisphere.* Polar Geography 23(2): 132-154.

ZIMOV, S. A., VOROPAEV, Y. V., SEMILETOV, I. P., DAVIDOV, S. P., PROSIANNIKOV, S. F., CHAPIN, F. S. III, CHAPIN, M. C., TRUMBORE, S., TYLER, S., 1997: *North Siberian lakes: a methane source fueled by Pleistocene carbon*, Science 277: 800-802.

7 Appendix

Table 7-1: List of surveyed plant species at investigation sites.

	Micro-site P I	Micro-site P II	Micro-site P III	Micro-site P IV	Micro-site P V
shrubs		1			
herbs	10	2	20	15	7.0
mosses	100	100	100	100	100.0
lichens		1			1.9
Salix glauca or S. reptans		1			
Pyrola rotundifolia					0.8
Carex chordorrhiza	8		10	10	
Carex concolor	0.1	2	10	3	5.2
Comarum palustre	1		5	3	
Pedicularis sudetica	0.1		0.1		
Lagotis		0.1			
Polygonum viviparum		0.1			
Poa arctica		0.1			0.1
Dryas punctata		0.1			1.5
Hierochloe pauciflora		0.1			
Equisetum arvense		0.1			
Luzula tundricola		0.1			
Lloydia serotina					0.1
Astragalus frigidus					0.7
Luzula confusa					0.1
Sacifraga punctata					0.8
Parrya nudicaulis					0.1
Saxifraga spinulosa					0.1
Saussurea sp.					0.3
Draba pilosa					0.1
Drepanocladus revolvens	99		95		
Drepanocladus cf. vernicosus	0.1				
Meesia triquetra	0.5		1		
Calliergon giganteum	0.1		0.1	0.1	
Aulacomnium turgidum	1	2	5		0.9
Polytrichum cf. alpinum					1.8
Hylocomium splendens		85			54.8
Tomentypnum nitens		10			
Campylium stellatum		5			
Rhytidium rugosum					40.4
Scorpidium scorpidioides				100	
Peltigera sp.		1			0.8
Dactylina arctica					0.5
Stereocaulon sp.					1.4
Cetraria laevigata					0.3
Cetraria islandica					0.3

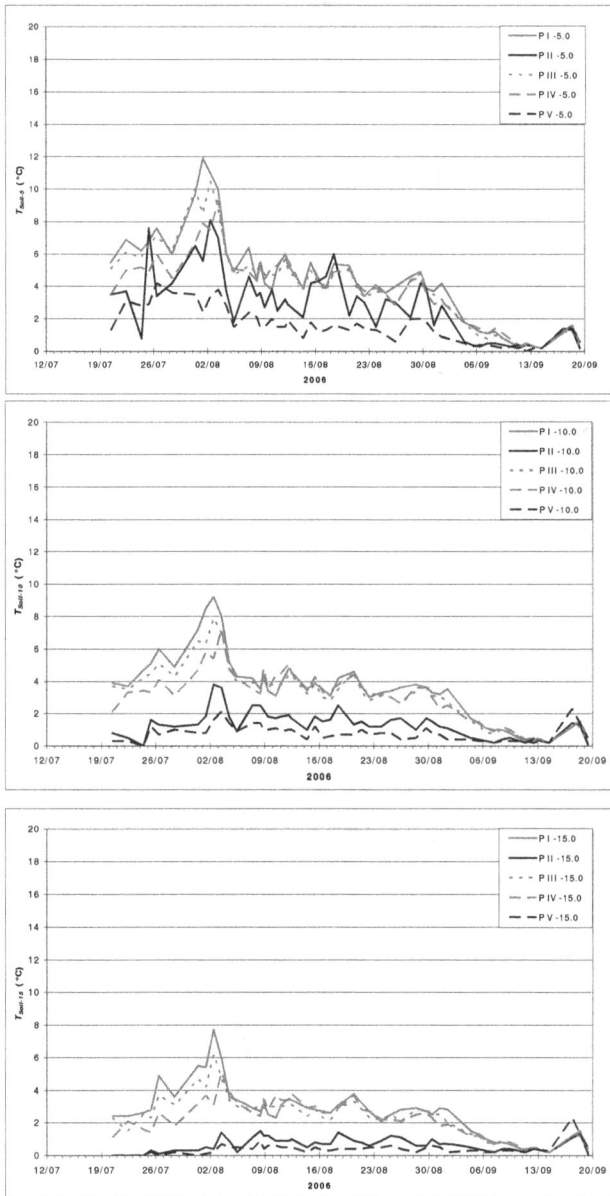

Figure 7-1: Soil temperatures at levels from 5 cm to 15 cm below ground.

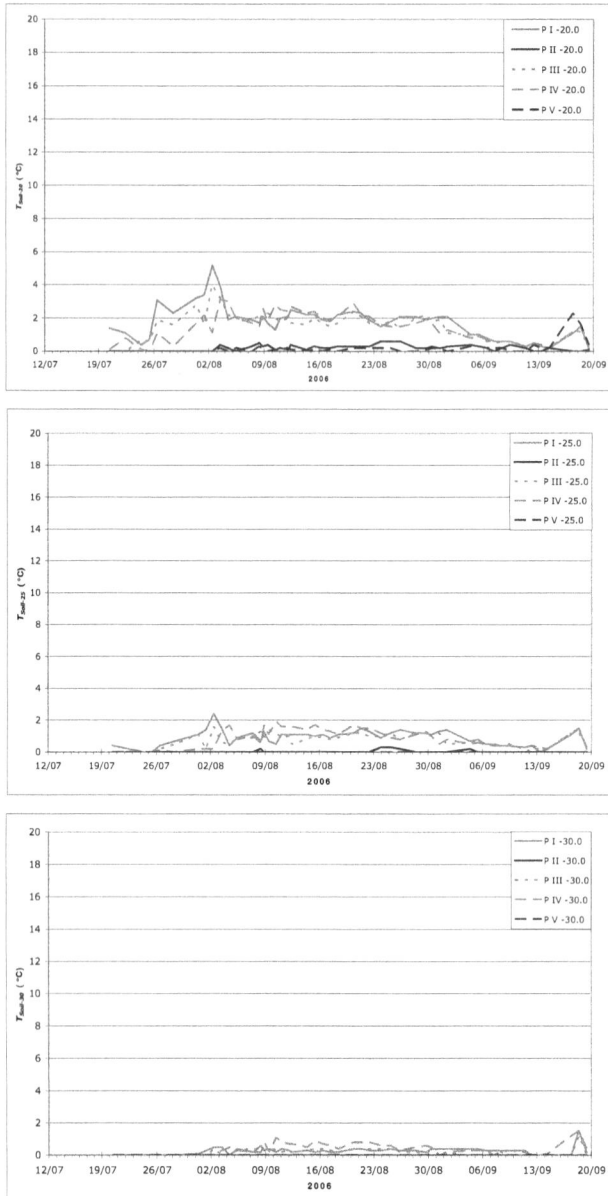

Figure 7-2:   Soil temperatures at levels from 20 cm to 30 cm below ground.

www.ingramcontent.com/pod-product-compliance
Lightning Source LLC
Chambersburg PA
CBHW061612220326
41598CB00024BC/3566